THE MANX ELECTRIC RAILWAY

Barry Edwards

B & C Publications

Copyright 1998, Barry Edwards

Published By
B & C Publications
81 Aylsham Drive
Ickenham
Middlesex UB10 8TL

Printed in England by The Amadeus Press, Huddersfield.

ISBN 0 9527756 2 X

All photographs not otherwise credited were taken by the Author.

All rights reserved.
No part of this publication may be reproduced, stored in a retrieval system, transmitted in any form or by any means, electronic, mechanical or photocopied, recorded or otherwise, without the prior written permission of the publisher.

Several of the photographs reproduced in this book are the only known views of particular tramcars. Examples of this are Nos 10, 13 and 24. The author, who can be contacted at the publishers address, would be pleased to hear from anyone who has photographs of the Manx Electric Railway that may fill these gaps.

Title page: **Car 20 trundles south on a cold May day in 1997. Snow had fallen on the higher mountains of the Island and the whitened slopes of North Barrule form a backdrop to the properties of Glen Mona seen to the left of the tram. Dhoon will be the next stop for the car on its journey to Derby Castle.**

BIBLIOGRAPHY

Industrial Archaeology of the Isle of Man: Various; Manx Experience, 1993.
The Isle of Man Railway, Volumes 1, 2 and 3: James I. C. Boyd; Oakwood Press, 1993/4/6.
Isle of Man Railways Fleet List: Barry Edwards; Author, 1994.
The Manx Experience, Eighth Edition: Gordon N. Kniveton; Manx Experience, 1994.
Manx Transport Kaleidoscope: Mike Goodwyn; MER Society, 1995.
Manx Transport Review (various issues); MER Society.
Manx Transport Systems: William Lambden; The Omnibus Society, 1964.
Portrait of the Isle of Man: E. H. Stenning; Robert Hale, 1975.
Rails in the Isle of Man: Robert Hendry; Midland Publishing, 1993.
The Railways and Tramways of the Isle of Man: Barry Edwards; OPC, 1993.
Snaefell Mountain Railway 1895-1995: Barry Edwards; Midland Publishing, 1995.
Isle of Man Tramways: F.K. Pearson; David & Charles 1970
100 Years of the Manx Electric Railway: F.K. Pearson; Leading Edge 1993.
Douglas Centenary: Gordon N. Kniveton; Manx Experience 1996.
Isle of Man Steam Railway: Barry Edwards; B & C Publications 1996
Isle of Man Railways Fleet List (second Edition): Barry Edwards; B & C Publications 1997.
Manx Electric: A.M. Goodwyn; Platform 5 1993.
The Manx Electric Railway: G.K. Kniveton & A. Scarfe; Manx Experience 1994.
The Dumbell Affair: Connery Chappel; Stephenson 1981.
Manx Electric Railway Album: Dr R. Preston Hendry & R. Powell Hendry; Hillside Publishing 1978.

INTRODUCTION

The Manx Electric Railway was born just over a century ago on an Island whose parliament was already some 900 years old. The parliament, or Tynwald as it is more commonly known, is now the longest running continuous parliament in the world. Its members take their seats in the House of Keys.

Situated in the middle of the British Isles, the Isle of Man is about 30 miles long and 10 miles wide, with a total area of 227 square miles bordered by over 100 miles of coastline. The 1991 census gives a population of 69,788, of which 44% live in Douglas and Onchan. The Island passed into the ownership of the British Crown on 11 July 1765, King George III becoming King of Man. The present Queen is Lord of Man.

The Island, which is probably best known for its annual TT (Tourist Trophy) motorbike races, offers something for everyone, with mountain and coastal scenery, heritage sites such as the Laxey Wheel and mines and Peel Castle, the hustle and bustle of Douglas with its various forms of entertainment and the peace and quiet of the Sound with its resident seal population.

The Manx Electric Railway runs through some of the most spectacular landscape, in particular between Laxey and Ramsey. It was the brainchild of Messrs Bruce and Saunderson and one wonders whether they ever imagined that it would survive to celebrate its centenary and beyond. That it has is in no small way a tribute to all those who over the years have been employed on the railway, providing the Island with what is now one of its biggest attractions. It all could so easily have been lost several times over, way back in 1900 with the collapse of Dumbell's bank, in the late 1950's just before the Government took over and in the mid 1970's when the northern section did close for one season. Thankfully it has survived, intact apart from the tramcars lost in the Laxey fire of 1930. Cars 1 and 2 are now the oldest electric tramcars in the world still working on their original line.

This book would not have been possible without the generous assistance and support offered by the management and staff of Isle of Man Railways and my sincere thanks go to them. Thanks also to the staff of the Manx Museum in Douglas and to David Bailey, Hugh Davies, David Haynes, Peter Johnson, David Lloyd-Jones, Terry Russell, Graham Stacey, my father and my wife Carol for the part they have played.

Barry Edwards
Ickenham
Middlesex
February 1998

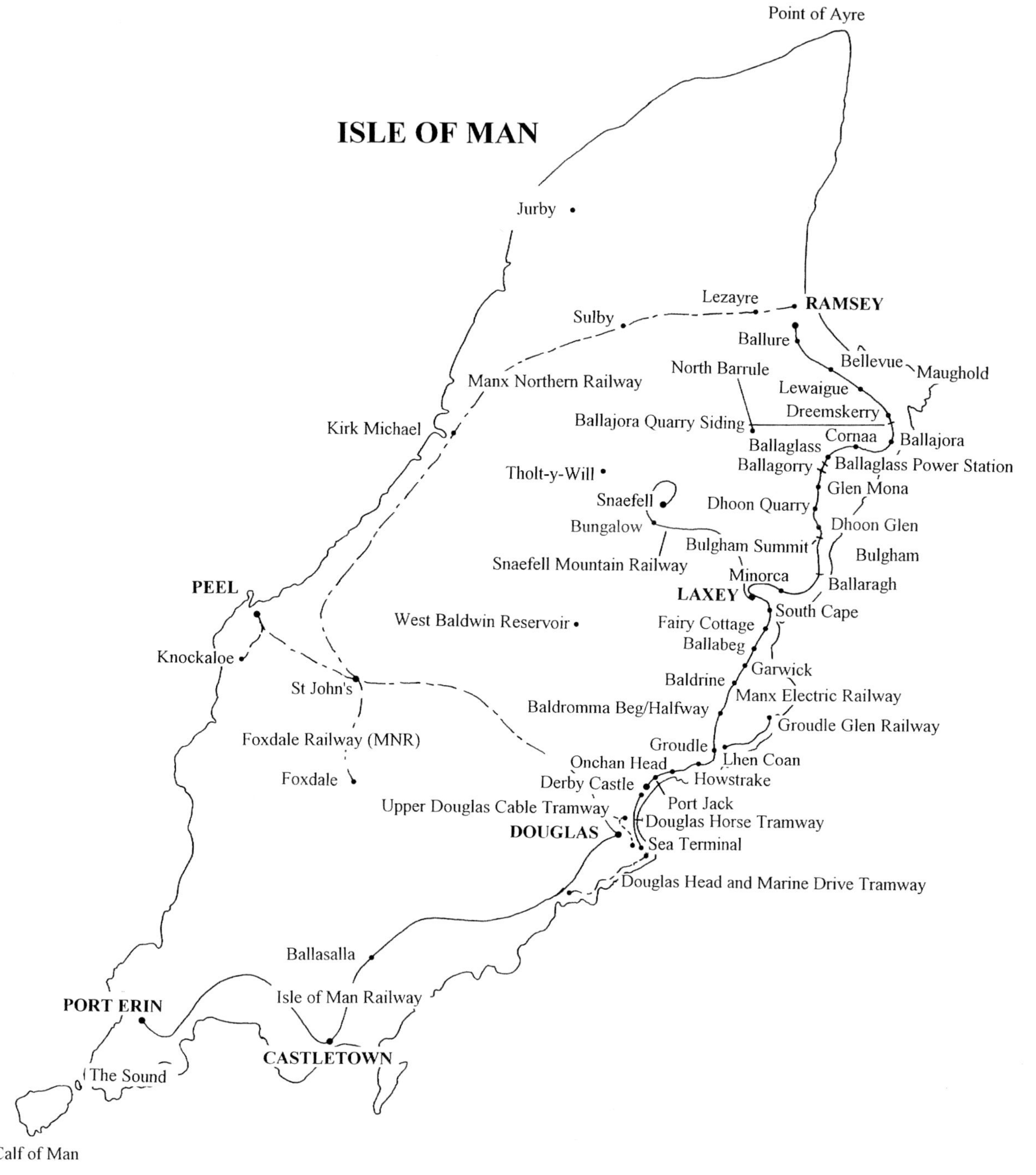

ISLE OF MAN
Point of Ayre
Jurby
Lezayre
RAMSEY
Sulby
Ballure
Bellevue
Maughold
North Barrule
Lewaigue
Manx Northern Railway
Dreemskerry
Ballajora Quarry Siding
Cornaa
Ballajora
Kirk Michael
Ballaglass
Ballagorry
Ballaglass Power Station
Tholt-y-Will
Glen Mona
Snaefell
Dhoon Quarry
Bungalow
Dhoon Glen
Bulgham Summit
Bulgham
Snaefell Mountain Railway
Minorca
Ballaragh
PEEL
LAXEY
South Cape
West Baldwin Reservoir
Fairy Cottage
Ballabeg
Knockaloe
Garwick
Baldrine
St John's
Manx Electric Railway
Baldromma Beg/Halfway
Groudle Glen Railway
Foxdale Railway (MNR)
Groudle
Lhen Coan
Foxdale
Onchan Head
Howstrake
Derby Castle
Port Jack
Upper Douglas Cable Tramway
Douglas Horse Tramway
DOUGLAS
Sea Terminal
Douglas Head and Marine Drive Tramway
Ballasalla
PORT ERIN
Isle of Man Railway
The Sound
CASTLETOWN
Calf of Man

THE MANX ELECTRIC RAILWAY

The Isle of Man had entered the railway era in the early 1870's when the Isle of Man Railway Company's steam operated lines opened from Douglas to Peel in 1873 and to Port Erin in 1874. Ramsey got its first railway in September 1879 when the Manx Northern Railway opened between St John's and Ramsey, this line following the west coast of the Island before turning inland via Sulby and Lezayre.

The latter part of the 1880's was to see a proposal, backed in part by the Island-based Dumbell's Bank, to build a holiday camp at Howstrake with a new road incorporating a tramway linking it with Douglas.

Alexander Bruce, General Manager of Dumbell's Bank and another gentleman, Frederick Saunderson, who had been involved in railway engineering in Ireland before moving to the Island in 1865, agreed in principle to purchase part of the Howstrake estate in 1889/90. The Howstrake Estate Act was passed in the House of Keys in March 1892 and included provision for a three-foot gauge single or double track tramway operated by steam, horse, electric or other traction. One wonders what other traction the members of the House of Keys had in mind.

Construction of the tramway started almost immediately and the first work included the filling-in of Port-e-Vack creek, the Derby Castle depot site.

Bruce was treating the construction and operation of the tramway as a testing ground for a line through to Laxey and was already looking at extending the line from Howstrake to Groudle. The lessee of the Groudle estate was persuaded to give up some of his land to allow the tramway to terminate opposite his new hotel, no doubt bringing additional custom for the hotel with it. The hotel obtained its licence on 4 July 1893 and opened the following day.

Meanwhile, such was the determination of Bruce to complete the line to Laxey that he had Saunderson survey the route as far as Baldromma Beg (Halfway) in advance of any announcement of his intentions. The Douglas and Laxey Coast Electric Tramway Co. was registered on 7 March 1893, to construct and operate a tramway between Douglas and Laxey.

Construction of the car sheds and power station at Derby Castle were advancing well and the first rails for the Groudle line arrived in May 1893.

The Derby Castle power station contained two 6ft diameter Galloway

The MER operated a door to door parcels service, based around the principal stations on the line. This Bedford vehicle started life as one of the Bungalow to Tholt-y-Will coaches and was converted to a parcels van in 1953. It is seen here parked near the Derby Castle terminus. Withdrawn from service in 1957, it was sold for scrap in 1962.

Author's collection

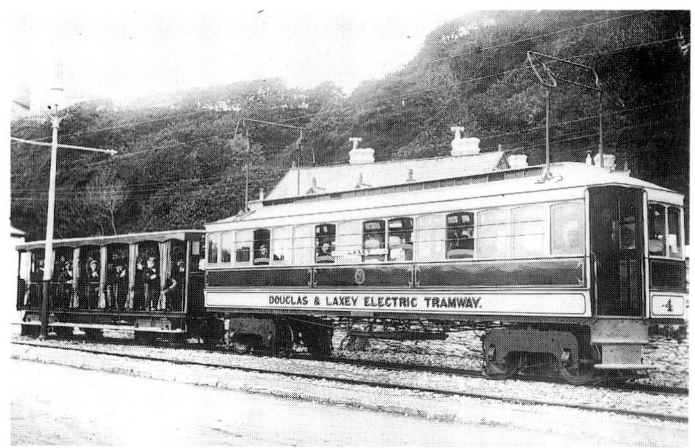

Top left: **Car 4 was one of the Laxey fire victims and few pictures exist. Seen here at Derby Castle about 1900, it displays the modified version of the Hopkinson bow collector; the originals were similar to those still fitted to the Snaefell cars. Of note are the Milnes plate bogies which had come from car 16 in 1899 and the 'Laxey' destination board across the middle side windows.**

Isle of Man Railways

Left: **Horse tramway car 32 hauled by** *Chloe* **heads away from Derby Castle bound for the Sea Terminal, while cars 43 and 37 await their next duties. Manx Electric car 2 and trailer 13 stand in the background along with a vintage Douglas Corporation bus and several historic cars. Viewed from the top of the double-deck horse tram (No.18) this area was once covered by the ornate IOMT&EP Co. canopy.**

Left: **Derby Castle is also the location for this fine picture of car 8, another consumed by the flames at Laxey in 1930. The turreted buildings behind the car and trailers are part of the adjacent entertainment complex.**

Isle of Man Railways

boilers, each 20ft long, with a working pressure of 120lbs per square inch. These supplied two 90 horsepower Galloway vertical compound engines, each running at 150 rpm. 9'0" diameter flywheels and leather belts drove Mather and Platt dynamos at 700 rpm producing 500 volts at 100 amps.

The first three trams arrived from G F Milnes of Birkenhead and the trackwork and overhead were complete but current collection problems prevented the line from opening. To resolve this, the overhead wire was raised at each pole to just clear of the bow collectors, the sag in the wire allowing the bow at the other end of the car to collect current.

It is likely that the first passengers were carried on 26 or 28 August but the line was not officially opened until 7 September 1893, operating for just 19 days that year during which 20,000 passengers were carried and around 1690 tram miles covered.

The Douglas and Laxey Electric Tramway Act was passed by Tynwald on 17 November 1893 allowing the compulsary purchase of land north of Baldromma Beg to Laxey, the work on building the line to be completed within two years. This time motive power should be animal or electric, other forms only being allowed if agreed by the Highways Agency. The maximimum gradient should be 1 in 20 and the tightest curves 90ft radius.

The company changed its name to Isle of Man Tramways and Electric Power Company Ltd (IOMT&EP Co.) after a successful bid of £38,000 for the Douglas Horse Tramway, which was handed over on 1 May 1894. The Groudle line, which was doubled in the early part of 1894, reopened on 12 May and was carrying 78,000 passengers a week by July. The three-span Groudle viaduct was built over Lhen Coan to carry the line northwards to Laxey.

Six new cars were ordered from G F Milnes and, following inspection by Messrs Rich and Cardew from the mainland Board of Trade, the extension to Laxey was opened on 28 July 1894. The Laxey terminus at this time was adjacent to the present car sheds. The new tramway attracted considerable press coverage as far afield as the United States, with the heavy engineering undertaken during construction affording particular attention.

Electrical supply for the extension was provided by a third boiler, engine and dynamo-set at Derby Castle and by a new steam power station at Laxey. The latter was built just above the river, on the steeper south side slopes of the valley, below the railway ledge. An additional shed to hold eight tramcars was built at Derby Castle.

Car 2 proudly displays its Manx Electric Railway centenary headboard, before following sister car 1 to Groudle with the centenary specials on 7 September 1993. The superb condition of these historic cars is very evident from this picture.

The words 'Electric Power' had been included in the company title as it was intended to provide the Island's public with electricity. The Douglas Bay Hotel (now demolished) with 250 electric lights became the first customer in 1894 and the Derby Castle Opera house joined the ever increasing list soon after.

The arrival of the tramway in Laxey had generated much excitement, to the extent that the IOMT&EP Co. was invited to send one of its directors to open a church bazaar in late 1894. During his speech Dr Farrell announced

Although undated this photograph of Derby Castle station was taken about 1900. The elaborate canopy was built over the horse tramway terminus in 1896 and was finally demolished in 1980. The third tramcar is one of the 1895 built 10-13 series power cars, all of which were withdrawn by 1903/4. The impressive buildings of the adjacent leisure complex are clearly visible.

Manx National Heritage

Opposite Page: The MER has only ever had two overbridges. One at Ballagorry still survives, the other at Summerland was demolished recently. The author used it as a vantage point to record car 21 shunting with trailer 40, while in the background car 9 awaits departure. The bases of the flagpoles are all that remains of the former canopy which would have covered the collection of horse cars awaiting their next duties.

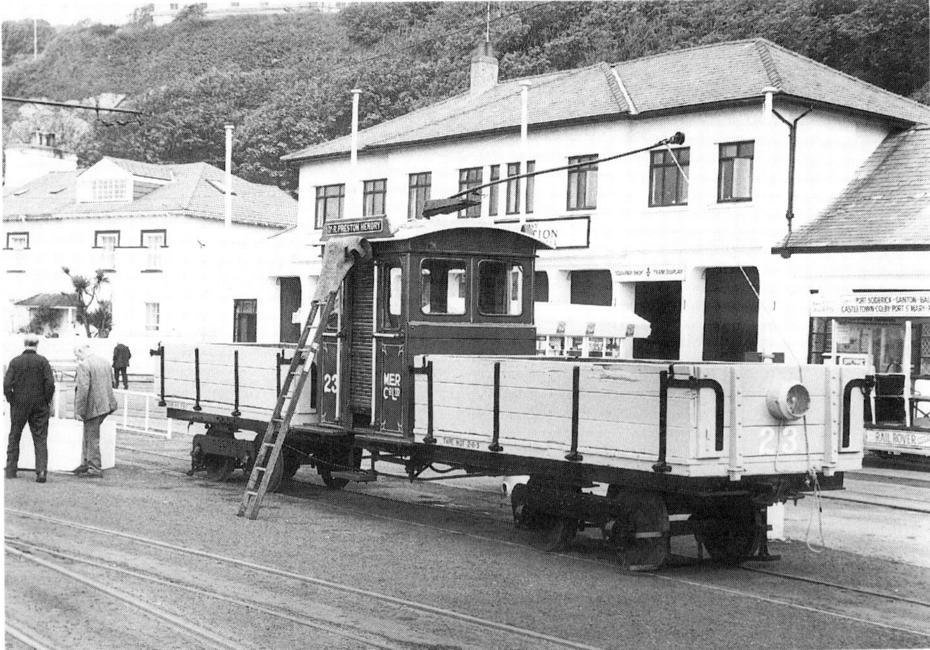

Following the death of Isle of Man Railway Society chairman, Dr R Preston Hendry, the society named locomotive 23 in his memory. Shortly before the ceremony at Derby Castle on 25 May 1992, 23 displays the soon to be officially unveiled nameplate.

a plan to build a tramway to the top of Snaefell, the Island's highest mountain.

Several earlier attempts had been made at this including one by Mr G B Fell, the inventor of the Fell rail system and he became involved in the new project. Electric traction was chosen and the line was built and operational in just seven months.

The full story of the mountain railway is available in the author's earlier book 'Snaefell Mountain Railway 1895-1995', published by Midland Publishing in 1995.

Bruce approached the Douglas Commissioners in 1894 about plans to provide a tramway to upper Douglas. The idea sat on the shelf for a while but was revived in 1895 on the basis that the IOMT& EP Co. would build a tramway to upper Douglas, double the remainder of the horse tramway and extend it to the steam railway station in North Quay, electrify it, and provide lighting for the Promenades and Victoria Street. Ten percent of the takings from the horse tramway would be paid to the Commissioners.

Bruce met with the Commissioners on 30 May 1895 to detail his proposals but omitted the electrification of the horse tramway. The ten per cent had increased to fifteen and in addition the horse tramway shed at the bottom of Summerhill would be removed and replaced by the one still in use at Derby Castle.

At the end of July the Commissioners accepted these revised proposals and the Upper Douglas Tramway Act was passed by Tynwald on 8 November 1895. The tramway opened on 15 August 1896.

The Corporation of Douglas was formed in February 1896, taking over from the Town Commissioners and another approach was made by the IOMT&EP Co. with a plan to double and electrify the horse tramway allowing the Manx Electric to operate through to the Sea Terminal. Branch lines to other parts of Douglas were also considered.

Still not content, Bruce launched plans for a further extension of the line from Laxey to Ramsey. A London based group had attempted to obtain an Act of Tynwald for a similar proposal in April 1896 and in October the IOMT&EP Co. petitioned Tynwald for leave to present its own bill.

The London group withdrew its proposal in December 1896 and the IOMT&EP Co. also withdrew, to allow time to present an improved plan avoiding the construction of a third station in Laxey. The coastal line would have terminated either side of Glen Roy and the Snaefell terminus was by now adjacent to Ham and Egg Terrace, having moved from opposite the car sheds earlier in the year. Various ideas were discussed, the main hurdle being

Top left: **The old top shed at Derby Castle was demolished in late 1996 to make way for a modern structure. Taken during demolition progress this picture clearly shows the reason for replacement.**

David Lloyd-Jones

Top right: **The replacement shed provides a much improved environment for the MER's historic fleet. This picture taken in May 1997 shows the shed nearing completion, the all important sprinkler system clearly visible in the roof area.**

Left: **This line-up in the old shed was photographed in August 1996. Cars 18 and 32 stand with trailer 49 and car 26 posing as trailer 42, the number it carried before motorisation by the railway in 1903.**

the crossing of Glen Roy to allow through running.

Following a new petition from the IOMT&EP Co. in March 1897, the Douglas and Laxey Tramway (Extension to Ramsey) Act was passed on 13 May. A viaduct would also be required at Ballure Glen on the outskirts of Ramsey, two years being allowed for construction of the whole line. Work started officially in the November although some preparatory work had been carried out in August. 250,000 tons of spoil were removed and 60,000 tons of ballast used in the construction.

A final attempt at electrification of the horse tramway and an extension to the steam railway station was made in early 1897. After considerable discussion and visits to similar undertakings, the go-ahead was given on 31 August 1897 but an outcry by local residents against the scheme ensued and, at a council meeting on 22 November, a letter from the IOMT&EP Co was read out withdrawing the proposal.

The current rustic ticket office at Derby Castle was completed in 1897 and a further extension to the car sheds was completed the same year.

An Andrew Barclay 0-4-0 saddle tank locomotive (Works No 713 of 1892) and forty-five ballast wagons were purchased to assist with construction of the Ramsey extension. The Isle of Man Railway's locomotive No.2 *Derby* and Manx Northern Railway No.1 *Ramsey*, with wagons 20, 21, 24 and

29 were also borrowed. On completion of the line, the Andrew Barclay locomotive was sold to Douglas Corporation for use at West Baldwin reservoir, where it was christened *Injebreck*.

Perhaps the most spectacular construction work was at Bulgham where the line is some 600ft above sea level. The existing road was closed and moved inland by 20ft, the tramway being laid roughly on the site of the original road.

The Snaefell line was extended a second time after a change of mind by the Highways Board and Church authorities in Laxey, allowing the present station to be built. The Snaefell line's terminus building was moved from its site in Ham and Egg Terrace to the new station and

replaced by an elaborate new 140ft x 40ft building in 1899. Sadly this was destroyed by fire on 24 September 1917.

The Dhoon granite quarry was provided with sidings and a weighbridge. The remains of the sidings can still be seen.

A new power station was built at Ballaglass to provide power for the northern section.

In early 1898, there were rumours of a proposal by the IOMT&EP Co. to build

Cars 5, 9 and 19 pose in their new surroundings in August 1997. Car 9 has had its side and end display panels changed in readiness for the 'Steam 125' celebrations of 1998. Final completion of the shed and track has taken place over the recent winter.

Left: Former Lisbon Tramways car 360 was purchased in mid-1996. Seen here in the company of MER cars 1 and 22, the modern looking (by comparison) Portuguese car is itself over 90 years old.

Bottom left: The Laxey shed fire in 1930 destroyed four power cars which were never replaced and this is the only known picture of car 24, seen here at Derby Castle depot in the company of trailers 57 and 58 in June 1922.

Late Reverend Bertram J. Kelly

Below: A general view of the lower car shed at Derby Castle in July 1950. Trailer 51 is just visible on the far left track, then vans 4 and 16 and an unidentified example keep company with cars 16 and 15 and an MER lorry. The lorry carries 'Manx Electric Co' and the number '6' on the cabside.

P W Bradley

a new line from Douglas to St. John's and take over and electrify the Manx Northern line from St. John's to Ramsey, thus creating a circular tramway round the northern half of the Island.

The Laxey to Ramsey extension was complete as far as Ballure by July 1898, apart from a short section of single line at Ballagorry Cutting. Messrs Rich and Cardew visited again and raised some concerns about the single line section. These issues were quickly resolved and the line opened to Ballure on 2 August 1898. The Lieutenant Governor attended the opening ceremony, the first such recognition of the success of the tramway and public traffic began later the same day. A car shed was built at Ballure and moved to the present site in Parsonage Road when the final section of the line was completed.

The site of the Ramsey terminus was the subject of some disagreement with the Town Commissioners; Bruce wanted to terminate adjacent to the Palace Concert Hall, the Commissioners preferred the Pavilion. Bruce ordered four Bonner road/rail wagons and purchased additional land at the Pavilion site to allow the operation of the wagons to and from the harbour.

A further debate among the Commissioners resulted in an agreement that the line could go to the Palace site, with possible extensions to the Harbour and Queens Pier.

The two 80ft spans of Ballure Viaduct

were fabricated by Francis Morton & Co. of Garston, Liverpool and the track at Ballagorry cutting was doubled. More new tramcars arrived for the Ramsey service.

The service resumed for the 1899 season as far as Ballure on 17 June, with an empty car reaching the Ramsey terminus on the afternoon of 3 July. A mail service along the length of the line started on 11 July. Messrs Cardew and Rich inspected the extension and the track at Ballagorry cutting and found all to be well.

Local advertising announced that the public opening would take place on

Locomotive 23 was built by the IOMT&EP Co in 1900 and borrowed bogies from car 17. A very elegant little locomotive, it was damaged in a derailment in 1914 and stored as a result. It remained on the books until 1922 but was not actually disposed of. During 1926 it was rebuilt onto a new longer chassis and the original cab was sandwiched between two 6ton wagon bodies; it now borrowed bogies from car 33. It operated until 1944 when it again went into store. It is currently preserved and ventures out for special occasions.

Isle of Man Railways

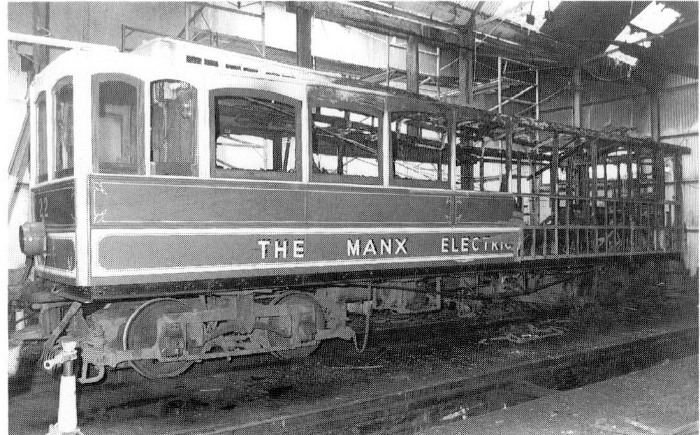

Top left: **Car 9 poses for the camera outside the lower shed at Derby Castle. Also visible are van 3, open wagon 1, power car 18 and trailer 50. Trailer 50 has seen little use recently and now lies stored at Laxey, while car 9 became the illuminated tram in 1993.**

Hugh Ballantyne

Top right: **On the night of 30 September 1990, car 22 caught fire while parked for the night in the lower car shed at Derby Castle depot. This was the scene that greeted staff when they arrived for work the following morning. The damage done to the recently refurbished shed roof is also evident in this view.**

Island Photographics

Left: **The conversion of trailer 56 for use by disabled passengers, which included increasing the overall body width, was carried out by the staff at Derby Castle depot. During the latter stages of conversion it is seen here on 8 February 1995 in the lower car shed, with painting well underway.**

Saturday 22 July 1899 but it actually opened on the following Monday 24 July. A half hourly service starting at 07.30 and finishing at 21.00 would operate with a return fare between Derby Castle and Ramsey of 3s 6d reduced to 2s 6d after 16.00.

The sale of tickets was suspended on the August Bank Holiday as the line was full to capacity, a situation that occurred on a regular basis. On one of these days, 16 August 1899, 4,500 people travelled from Derby Castle to Ramsey. The Ramsey station building, which had previously been at Ballure, was re-erected and ready for use by 25 August.

Three Bonner wagons were delivered (not four as ordered) from the Bonner Wagon Co. of Toledo, Ohio, on 1 September 1899. These were ingenious wagons, with the rail wheels lower than the road ones and the transfer being completed by shunting the wagon into a siding with ramps either side of the track, raising the wagon to release the rail wheels. The wagon was then hauled away on its road wheels.

The wagons were to be used for transporting coal from the quayside at Ramsey to Ballaglass Power Station and stone from Dhoon Quarry to Ramsey Harbour. Road/rail transfer points were constructed at Ramsey station, Laxey and Derby Castle but complaints about the noise at the former resulted in a portable structure being used in Queens Drive instead.

The power supply for the entire system

was by now supported by three battery houses, one at Groudle, one at Ballaglass and one on the Snaefell line. Power would need to be supplied to these all year round, which would mean keeping at least one of the steam power stations open. Deep in the Laxey Valley, about half a mile down from the steam power station, the river is flowing with some vigour and this point was therefore chosen as the site for a new water turbine power station. Two 720rpm turbines were installed, both coupled to an Electric Construction Co. bipolar dynamo which produced 520 volts at 120 amps. The power was transferred via fifteen overhead cables to the steam power station and onwards from there by existing underground cables to the battery houses.

A 40'0" long, 4'6" high concrete dam was built across the river, the water being taken via two tanks to the

An undated view inside the old top shed at Derby Castle depot. 1894 power car 7, still with its twin windscreens, is in the company of power car 15 and trailers 46 and 55.

Author's collection

turbines. Considerable care was needed to avoid waste from the mines further up the valley getting into the turbines.

Alexander Bruce carried out what was to be his last public duty on 17 December 1899 when he started the turbines for the first time. Testing of the system commenced and the power station provided a regular service from 27 December.

The tramway now had steam driven power stations at Derby Castle, Laxey, Ballaglass and Snaefell, the Laxey turbines and the three battery houses.

The new century started badly for the

Far left: Shortly after leaving Derby Castle the cars pass the entrance to the depot and then begin the climb to Port Jack and Onchan Head. Car 21 with trailer has just negotiated the depot entrance pointwork and will now accelerate away towards Laxey. The concrete slope arrangement on the other side of the road is the seaward slope/staircase of Summerland's former footbridge, which until demolished was the second of the line's overbridges.

Left: An unusual view, from Imperial Terrace in Onchan, of a winter saloon as it descends from Port Jack to Derby Castle. The buildings on the left are part of the Port Jack parade.

Opposite page: Car 6 and trailer ease round the sharp curve between Onchan Head and Port Jack while working a Derby Castle bound service. The pair have just passed the entrance to the former Douglas Bay Hotel. The 'White City' sign was for the now closed amusement park.

Left: Car 2 hauls Royal Trailer 59 past Port Jack while working a special to Ramsey in May 1989. Trailer 59 was built as a 4-wheeler but soon received bogies to improve the ride quality. It now forms part of the display in the Ramsey visitor centre from where it can be removed to work on the line if required.

tramway. The supporting Dumbell's bank closed at 10.00 on Saturday 3 February 1900, causing financial chaos on the Island.

It will be recalled that Bruce was General Manager of Dumbell's and it transpired that the collapse was due to a loan made to the tramway company of £65,000, which it was said had come from England. This was in fact true but it had come via Dumbell's and the English bank now wanted its money back. The IOMT&EP Co owed Dumbell's in the region of £150,000 but was unable to meet the debt which in turn prevented repayment to the English bank.

Warrants for the arrest of the directors

of Dumbell's and of the IOMT&EP Co were issued, including one for Bruce. He, however, was now very ill and confined to bed and the High Bailiff accepted that his arrest could prove fatal.

The IOMT&EP Co annual general meeting held in May 1900 was faced with bad news. Despite vigorous attempts to secure a buyer for the entire undertaking, prospective purchasers were not interested because it was clear that a liquidation sale would be at a far lower price .

An order for six more Bonner wagons was rejected by the supplier in Ohio.

Several special meetings were

convened and the directors were replaced but to no avail, the liquidation being enforced at a court hearing on 11 July 1900. This all proved too much for Bruce who died just three days later. Sceptical locals were convinced that Bruce had in fact left the Island and that the coffin was filled with stones.

The appointed liquidator ordered the sale of all assets on 3 July 1901, nearly eighteen months after the Dumbell's collapse.

Offers were received from various interested parties ranging from £188,000 to £225,000 for all four lines. Douglas Corporation offered £50,000 for the horse and cable tramways, an offer which was eventually accepted on 25 September 1901. An offer was then received from a Manchester based syndicate, which turned out to be one merchant banker, of £250,000 for the Manx Electric and Snaefell lines. This was accepted, the sale was sanctioned by the Clerk of the Rolls on 5 September 1902 and completed on 14 November; the actual amount paid was £252,000.

This was to prove a profitable purchase, as a new company, The Manx Electric Railway Company, which had been incorporated in London on 12 November, purchased the two lines from the Manchester syndicate for £370,000, this sale being completed on 30 November 1902.

Meanwhile at Ballagorry a three-span wooden overbridge had been constructed in 1901. This was

Top: 'Ratchet' car 17 of 1898 trundles past Onchan Head station on its way to Derby Castle in July 1966. The now demolished White City amusement park roller coaster is visible behind the tram.

A D Packer

Middle: As the cars climb towards Howstrake the number of houses diminishes. On the outskirts of Onchan we see car 19 with trailer 47 heading for Laxey. A new overhead pole has been put in place to await the transfer of the wires before the old can be removed.

Opposite page: The longitudinal seating in the 1894 series cars led them to become known as the 'Tunnel cars'. Car 5 with trailer in tow has just negotiated the Howstrake curves and begun the descent into Derby Castle.

Bottom: Car 25 and trailer 55 glide down from Howstrake to Onchan Head on 2 August 1965. Having left Howstrake, the final mile and a half of the journey is downhill with the gradient varying between 1:23.6 and 1:141. The line drops from 258ft above sea level to just a few feet above at Derby Castle.

Author's collection

eventually replaced by the present concrete structure in the 1950's which, apart from the short-lived bridge at Summerland, remains the only overbridge on the line.

King Edward VII and Queen Alexandra travelled on the line from Derby Castle to Walpole Drive in Ramsey in trailer 59 on 25 August 1902, while all the turmoil of ownership was underway.

The first statutory meeting of the new company was held in Douglas and told by the permanent way superintendent that the trackwork between Derby Castle and Laxey was in need of major repair. 5,200 new sleepers, 120tons of rail and 20,000tons of ballast were required.

The Ramsey section was in much better order with only local reballasting and realignment required. The new company was also unhappy with the construction method at Bulgham and in January 1903 was successful in obtaining permission to close the adjacent road, carve back the rockface and move both the road and tramway inland, in order that the tramway would be running on a solid rock base. A total of 46,735tons of spoil were removed from the site during the operation.

In an attempt to reduce damage caused by road vehicles at crossings, heavier duty rail was laid at the busier sites. A longer term project was started to ease curves and to improve and repair such items as drains, hedges and platforms.

The new owners also appointed consultants to look at the electrical

systems on the line and advise on ways of improving and updating. The principal recommendation was to convert to AC power. This was duly accepted and the contract to carry out the conversion awarded to Witting, Eborall & Co. on 16 February 1903.

Laxey steam power station was chosen as the hub of the new system, the generating side being gradually rebuilt in the months up to July 1903. Electric sub-stations were commissioned at Derby Castle, Groudle, Ballaglass and Snaefell, where possible sharing buildings with the earlier generating equipment except at Derby Castle, where the steam engines were removed. New feeder systems were installed to carry the 7000 volt supply from Laxey to the other locations, in some cases the original underground feeder cables being re-used. Indeed some of these were still capable of service in the early 1970's, a quite remarkable lifespan.

The new generating plant was run for the first time in July 1903, the full system being tested in December.

Coal for the Laxey power station was brought up from the harbour by cart; later proposals for a short electric tramway and then for a branch of the main tramway never materialised.

The re-equipping of the tramcars to AC operation took place over the next couple of years, with the overall

Left: Car 20 runs through the outskirts of Onchan while working a Ramsey to Derby Castle service. The tight radius of the curve is well illustrated by the lean on the coach travelling in the opposite direction on the adjacent road.

Bottom left: Perhaps the most unusual station building on the line is at Howstrake. Built in 1910 by the adjacent Holiday Camp Company, it still stands.

Below: The four 1895 power cars, 10-13, were built at the same time as the Snaefell cars and were very similar in appearance. They had unglazed windows and only completed seven or eight years service, during which time few photographs appear to have been taken and, of these, most were at Groudle. It is likely that these cars were not popular with either passengers or staff and so spent a good deal of their time operating on the short stretch between Derby Castle and Groudle. Here we see car 11 at Groudle in 1895, while fitted with bow collectors of the type still in use on the Snaefell line.

Authors Collection

journey time on the line being reduced as a result.

The Manx Electric Railway concentrated on increasing the amount of freight traffic, converting passenger car 12 to carry livestock in 1903 and removing seats from trailers during the Spring to allow bags of manure to be carried. Special fares were offered to traders who sent their freight with the MER. The freight traffic also allowed more use of the line during the winter months.

The Post Office contract was extended during 1903, when the tramway conductors were trained as auxillary postmen and timed collections started from boxes along the line. Formal arrangements were agreed to handle the considerable quantities of mail posted at the summit of Snaefell.

The next annual general meeting of the company took place in London on 26 January 1904, attendees learning of a 20% increase in traffic and good financial results.

A new 150'6" x 40'0" shed capable of holding 16 tramcars was built at Laxey as was a large goods shed. Ramsey also received a goods shed.

1904 was a poorer season with passenger numbers dropping but this situation improved for 1905.

Freight traffic continued to grow, the variety of goods carried including cattle, horses, flour and considerable amounts of luggage for the Steam Packet Company. Regular stone trains

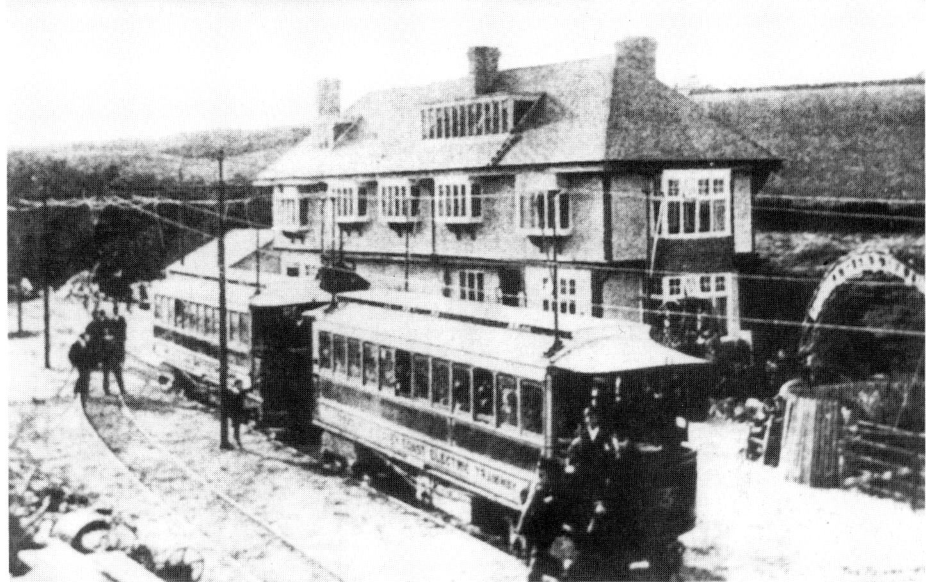

operated by locomotive 23 and usually two open wagons ran between Dhoon Quarry and Derby Castle where the stone was transferred to road vehicles for the remainder of the journey, often to the Sea Terminal for export.

Meanwhile the fortunes of the Manx Northern Railway were falling away, the line having been operated by the Isle of Man Railway with Government approval since February 1904. Although it seemed obvious that it should be taken over by the IOMR, the MER had other ideas and made a second bid for the line.

A takeover race ensued but by 18 April the MER had withdrawn. An Act of Tynwald allowing the IOMR takeover was passed on 24 May 1904, the final merger taking place on 19 April 1905.

Cars 1 and 3 stand at Groudle in 1893, before the line was extended to Laxey. The cars are on the reversing loop and the single track shunt is just visible behind the cluster of people to the left of car 1. The tracks appear to be closer to the hotel than they are today, indicating that they were moved away from the hotel when the line was extended, possibly to allow space for the curve onto the viaduct. The arched entrance is to Groudle Glen.

Author's collection

Over half a million passengers were carried during 1906, the growth in numbers being assisted by the opening of the new Snaefell Summit Hotel and the arrival of more new tramcars. Laxey Glen and gardens were purchased in the same year.

The tramway benefitted from serving a

Groudle Station, J. O. M.

Above: **The Lieutenant Governor unveiled this plaque at Groudle station after driving car 1 from Derby Castle on 7 September 1993.**

Top left: **Another view of Groudle with a 10-13 series car but this time No. 13. This picture was taken in 1898 after the cars had been fitted with trolley poles in place of the bows and before delivery of the 19-22 series winter saloons, when the trailer on the left numbered 20 would have been renumbered 37. Shunting for the return to Derby Castle is in progress. The power car has moved forward and then reversed over the crossover, the trailer will now be pushed away from the camera to allow 13 to reverse back and couple-up. The conductor already has the pole in position for the next move.**
A D Bailey Collection

Right: It is clear from this picture that the 10-13 series cars did on occasion work north of Groudle. Car 10 brings an unidentified trailer round the second of the 90 degree curves over Groudle viaduct towards the station. The number of onlookers perhaps indicates a special occasion. The small building was known as the Toll House, used until sometime in the 1920's to collect the toll from traffic using the private road to Baldromma Beg (Halfway); built in 1898 it was demolished in 1988.

Isle of Man Railways

Opposite page bottom left: **Winter saloon 21 at Groudle on 24 October 1997, coupled to an unidentified wagon carrying a generator. The power had been switched off over part of the route and in order to maintain through traffic the generator was put to good use. The pair are seen here preparing for departure while working the 10.00 from Derby Castle to Ramsey, the final electrical connections being made. The car's own trolley pole is tied down clear of the overhead wire.**

David Lloyd-Jones

number of glens along its route and at both Groudle and Garwick a small payment was received for each passenger entering the glens. Ballaglass was rented and later owned by the company and Dhoon was rented. Dhoon also boasted a large hotel and entertainment complex.

Connecting with Snaefell trams, a motor charabanc service was introduced between Bungalow and Tholt-y-Will for the 1907 season. Forty-two and a half thousand people

arrived on the Island for the August Bank Holiday. The queue at Derby Castle of intending passengers stretched as far as the foot of Summerhill, causing the horse tramway service to stop short of its usual terminus.

The company by now owned six licensed premises, at Douglas (Strathallan Hotel), Laxey Station Hotel, Bungalow, Snaefell, Tholt-y-Will and Dhoon plus the refreshment room and cafe at Laxey. Some of these had previously been let but were re-taken and a Hotels Manager appointed.

The MER arranged lighting along the Promenades for the 1911 Douglas carnival, resulting in a contract being placed by Douglas Corporation with a London street lighting company for a permanent display, the forerunner to the present day lights.

Frederick Saunderson, the second of the two pioneers of the railway, died on 9 July 1911 aged 71.

The 1914 season started very well with passenger numbers up on the previous year but the season came to an abrupt halt on 13 July with the declaration of war. Large internment camps at Douglas and Knockaloe benefitted the IOMR but brought little or no additional traffic to the MER.

The water turbine power station was almost sufficient to provide the required power during the war years.

Traffic levels began to improve again during the 1917 season but sadly the vast refreshment room at Laxey was destroyed by fire on 24 September and never replaced.

Further improvement in passenger

Top: **Winter saloon 20 approaches Groudle Viaduct round the first of two curves that take the line through 180 degrees. Contractor Mark Carine built the Viaduct with its three 20'0" spans for the extension to Laxey in 1894.**

Opposite page: **After crossing the Groudle viaduct the line follows the A11 road to Halfway where they meet the main A2 Douglas to Laxey road. Car 20 and trailer 42 pass the upper end of Groudle Glen, a good load of passengers enjoying their journey towards Laxey. Each of these four winter saloons (19-22) has completed over 4,000,000 revenue earning miles.**

Bottom: **Car 2 rounds the second of the Groudle curves on its way north in 1956. Despite its historical significance, this car was relegated to works duties for a time before being restored in the 1970's.**

RAS Marketing

levels in 1918, together with the sale of the charabancs and other redundant equipment failed to clear the war deficit. All transport services on the Island were halted on 4 July during a general strike fuelled by a threatened 33% increase in the price of bread (known locally as the 'bread strike').

Visitors began to flood back to the Island in 1919, the Snaefell line re-opened and the entire system produced an operating profit that wiped out the remaining war deficit.

Refurbishment of the power supply system that had been started before the war was resumed and continued into 1920/21. Passenger figures soared in 1920, the operating profit of £40,000 clearing all interest owed from recent borrowing. New rolling stock was considered and a three-car lean-to extension was built at Derby Castle depot.

Passenger figures dropped back a little in 1921 to just over 660,000 and increasing costs cut the overall profit margins. The winter service for 1924/5 still consisted of nine northbound and eight southbound workings, some starting or terminating at Laxey.

The Snaefell power station lost its steam engines in 1924, leaving just Ballaglass and Laxey so equipped.

The 1926 coal/general strike on the mainland caused a shortage of coal and hence generating problems for the tramway. Three car trains (two trailers) were tried but overheating controllers soon brought this to an end.

A once-daily express service, usually operated by one of the 19-22 series cars hauling a goods van, was introduced in 1928 and survived until 1937. It was timetabled to complete the journey to Ramsey in just 60 minutes.

The late 1920's and early 30's were not good times for the tramway. Motor bus services were introduced between Douglas and Ramsey via Laxey during the winter 1926/27, creating the first real competition the tramway had faced and thirty-two passengers were injured when a trailer car broke loose at Fairy Cottage in 1928.

The night of 5 April 1930 witnessed what probably remains the most serious incident the line has seen. Fire broke out inside the Laxey car sheds destroying 11 tramcars, including power cars 3, 4, 8 and 24, various goods vehicles and countless bits and pieces. A carelessly discarded cigarette end is thought to have started the fire. The depot was rebuilt and three replacement trailers were ordered from English Electric, these vehicles remaining the newest on the line.

A few months later, on the night of 17 September, a severe storm over the north-east and centre of the Island caused serious flooding in several areas and produced a build-up of between two and four thousand tons of debris

Left: Car 22 has just left Baldrine and is curving to the right (towards the left of the picture) to run parallel with the main Douglas, Laxey and Ramsey (A2) road. The station building is visible to the rear of the car. The mirror is provided to assist motorists to see approaching trams.

Bottom left: Garwick Glen station is little used these days, although a recent opening of the gardens of the converted mill led to a queue of intending passengers. Car 32 curves away from the station and heads for Baldrine and Derby Castle on 4 August 1957.

J H Price

MANX ELECTRIC
IN COLOUR

Car 16 and an unidentified trailer await departure from Derby Castle during the summer of 1964. The rustic ticket office was completed by the IOMT&EP Co. in 1897 and measures just 12'6" x 8'0". Lighting and heating are provided from the traction current.

Author's collection

Opposite page bottom right: **Car 27** with an unidentified trailer comes off the tight curve on the approach to Garwick station on 23 June 1926. Garwick station served the nearby glen which at its peak provided the railway with upwards of 100,000 passengers per year. The Glen was eventually sold to a private buyer and the water corn mill converted into a private dwelling. The station buildings were demolished in 1979.

LCGB Ken Nunn collection

The complicated overhead wiring in the entrance to Derby Castle depot requires special attention when shunting cars. Watched by a member of the depot staff, car 26 eases its trailer into the upper car shed during 1964. The 'Electric Railway' sign on the cliff behind the depot was refurbished for the line's centenary in 1993.

Author's collection

The Manx Electric had its first illuminated car as part of the centenary celebrations. Fitted with around 1600 individual bulbs the work was supported by the Seaton Tramway. Standing inside the lower sheds on 20 April 1993, car 9 keeps company with one of the winter saloons undergoing overhaul.

Winter saloon 21 of 1899 with trailer 47 bring a good load of holidaymakers back to Douglas at the end of a busy day. The Douglas Bay Hotel, which was destroyed by fire in 1988 and has since been demolished, is in the background.

Car 6 had its twin front windscreens restored following a collision in July 1991. In immaculate condition, it is seen here hauling the 'disabled' trailer 56 round the curve above Port Jack towards Onchan Head. At this point the passengers have a superb view across Douglas Bay.

Peter Johnson

Locomotive 23, with two wagons, trundles towards Groudle during September 1993. The locomotive was rebuilt to this condition in 1926 after the original was damaged in a derailment in 1914.

Top: Groudle is the first major stop, the station serving the nearby Glen and its narrow gauge railway. Winter saloon 19 and trailer 47 pause to allow passengers to alight, while working a Laxey to Douglas service.

Top: **During the summer an evening service runs to and from Groudle, now usually operated by the illuminated tram. Car 7 is seen here before the removal of its twin windscreens, awaiting departure for Derby Castle.**

P W Gray/Colour-Rail/NG50

Opposite page bottom: **10-13 series car 12 stands on the northbound track at Groudle station sometime between 1895 and 1898. The impressively painted shelter to the left of the car dates from 1894 and is still in use today. The glen entrance is to the left of Dobie's tearooms, the building for which was originally at the Iron Pier in Douglas and moved to this site when the pier was demolished in 1893.**

Authors collection

Bottom: **Car 5 heads for Douglas in the latter part of the 1995 season. Of interest is the guard checking tickets whilst clinging to the side of the trailer, a practice that has now ceased. In the background is Laxey Harbour and on the hillside beyond can be seen the route of the MER between Laxey and Ramsey.**

Peter Johnson

1893 cars 1 and 2 stand outside Laxey shed during July 1997. These are the oldest electric tramcars in the world still operating on their original line. Despite a period in departmental use they have both been fully restored and, as is clear from this picture, are in superb condition. They both see regular service during the summer months and will no doubt continue to do so for many years to come.

Peter Johnson

Laxey station is an idyllic setting, which is somehow undisturbed by the arrival of one of the MER's vintage trams. 1899 built car 21 arrives from Ramsey in brilliant sunshine. The two reversing crossovers are visible in this view, the wider track to the left is that of the Snaefell line.

Perhaps the most spectacular stretch of the line is that at Bulgham. Car 19 and trailer have just negotiated the clifftops and pass Ballaragh as they begin their descent into Laxey. On the distant headland is the 75ft high Maughold lighthouse dating from 1914. Its 3 flashes every 30 seconds would be clearly visible from this spot as it has a 22 mile range.

It is not often that a 'Ratchet' car ventures north of Laxey. Here though we see car 18 heading for Ramsey in July 1997. The motorman has his hand firmly on the brake handle. The superb scenery through which the line passes north of Laxey is clearly evident from this picture.

Peter Johnson

1895 built car 10 was converted to a bogie freight trailer in 1926 and is now preserved. It is seen here parked in a siding at Ramsey having been allowed out of the car sheds. It is in generally good condition and perhaps one day will be fully restored to 1895 condition.

An unidentified winter saloon hauls trailer and van 13 away from Ramsey during the summer of 1964. This formation would have been relatively common at this time, the van conveying parcels for addresses in Laxey and Douglas.

Authors collection

behind the dam for the Laxey turbine power station. This meant that vast quantities of water took to the roads causing widespread damage to property as far down as the harbour and flooding the power station itself. It was late November before the Laxey plant was fully reinstated, the Ballaglass power station providing extra power while repairs were carried out.

That was not the end of the problem for the MER as a legal judgement placed all responsibility for repairs to roads and clearing rubbish from the river with the company. The continuing threat of road transport and the costs incurred by the fire and flood caused financial worry.

In order to reduce costs by eliminating the requirement to provide full overhaul facilities for the mountain railway cars at Laxey, a dual gauge siding was laid in Laxey station to allow the transfer of Snaefell cars onto 3'0" gauge bogies for transport to Derby Castle.

The Dhoon Glen Hotel complex was destroyed by fire on the evening of 3 April 1932 and never replaced.

A public electricity supply, controlled by the Isle of Man Electricity Board, was now available across the Island and by 1933 the MER supplies to the Onchan area and parts of Derby Castle had been taken over by the new company. The MER's last electricty customer was the Howstrake holiday camp.

This led the MER to look at their own

supply as the ageing system was costing around £5000 a year to operate and so negotiations began with the Isle of Man Electricty Board to supply the line via a network of new sub-stations.

An agreement was signed on 7 November 1934 and new sub-stations were installed at Derby Castle, Groudle, Laxey, Snaefell, Ballaglass and Bellevue. Laxey, Ballaglass and Bellevue are fed direct from the grid at 33,000 volts, the remainder at 6,600 volts from a purpose-built transformer station at Laxey. Fan-cooled mercury arc rectifiers are used at all sites. The original 1904 switchgear was re-used as the marble facings do not age.

The Laxey turbines were last used in April 1935, with the new system coming into full use on 1 May.

Ballabeg station building dates from 1905 and was rebuilt in 1992. This view taken in 1981 shows the building after removal of the post box which stood to its left.

The increasing use of private motor vehicles on the Island led to the installation of automatic level crossing lights at Halfway in 1934 and at Ballure and Ballabeg in 1936. These systems were trolley activated and have been updated over the years.

Most visitors to the Island arrived on a Saturday and so Sunday was the very important 'Publicity' day; quantities of leaflets and other material were widely available and purchasers of a two-day rover ticket received a free four-page pamphlet.

Monday was the busiest day, Tuesday

Top left: **Trailer 52 lost its bodywork in 1947 and became a permanent way flat. Seen here parked outside Laxey car shed, the remaining end panel has been painted yellow with elegant shaded numbers.**

Top right: **Trailer 53 and an unidentified sister idle their time away in the Laxey car sheds. Although currently stored, 53 was one of the original 1893 trailers and probably numbered 15 on delivery. Photographed in 1956.**

R A S Marketing

Left: **Standing in the headshunt of Laxey shed are cars 17 and 27 with tower van 12 and a tower wagon between them. In the distance is permanent way flat 21, with its Challenger cranes visible. Car 17 is currently stored but is clearly repairable and it is perhaps surprising how little has had to be taken from it to keep others in the fleet running.**

and Wednesday shared the use of the second day of the rover tickets and Thursday saw the arrival of the weekly excursion from Fleetwood; Friday and Saturday were reasonably quiet. The MER was kept up to date with visitor arrivals each Saturday, in order that it could plan for staffing needs and tramcar availability for the coming week.

The steep slopes behind the Derby Castle depot were used to advantage with the building of an illuminated sign high above the depot which proclaimed 'MER for Scenery'. Damaged by a gorse fire, it was later changed to 'Electric Railway For Scenery'.

The outbreak of war in 1939 led to another significant reduction in passenger numbers. The MER was utilised to convey POWs to work on farm sites around the Island while peat from Snaefell and the transfer of mine waste from Laxey to Ramsey, for onward movement by road to Jurby airfield for runway construction, provided a good deal of freight traffic. A special siding was laid at Laxey for this purpose.

War weary visitors flocked to the Island in 1945, giving the MER a £11,000 working profit. There was, however, a considerable amount of maintenance to catch up on after a general neglect during the war years. Boosted partly by the reopening of the Snaefell line, over one and a half million passengers were carried in 1946, allowing £15,000 from the profits to go towards the required repairs.

The Tholt-y-Will tours resumed in 1946 but the end of petrol rationing in the early 1950's created competition from other coach operators and the service was finally withdrawn in 1952.

A general report on public transport on the Island recognised the social importance of the year-round service on the MER but also acknowledged the increasing threat from motor bus competition which was giving cause for concern over the future of the Laxey to Ramsey section.

In 1949 Douglas Corporation sought to obtain a licence to operate buses to Groudle but this was rejected; the buses would only be allowed as far as White City. The service ran until 1957 and again from 1959.

The number of visitors to the Island dropped by 75,000 in 1950 with the obvious effect on passenger numbers, the winter Sunday service falling victim to low demand in 1951/52.

Rails and poles salvaged from the now closed Douglas Head and Marine Drive Tramway were acquired, the rails proving to be very noisy so no more were obtained but the poles continued to be re-used on the MER.

An operating loss of £850 was incurred

Inside Laxey sheds are cars 15, 29, 26 and 30. Trailer 50 and van 3 are just visible to the left of the picture. It is hoped that the cladding on this 150'6" x 40'0" shed will be replaced in the near future.

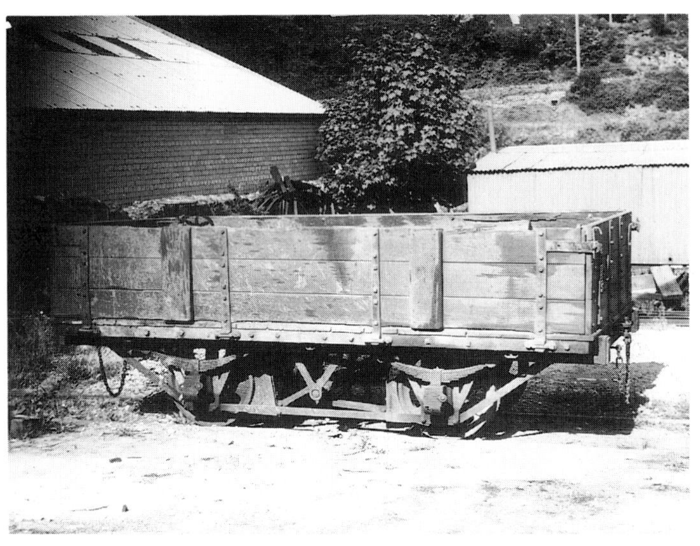

Above left: G F Milnes built this 6-ton open wagon in 1894. It became MER No.1 and is seen here at Laxey when just 56 years old.

P W Bradley

Above: During the reconstruction of the top shed at Derby Castle it was necessary to store all the tramcars elsewhere. Many of the service power cars were moved to Laxey sheds, the stored vehicles from there and many of the trailers being parked on the southbound running line between the depot and station. From right to left the cars are 46, 47, 44, 42, 61, 40, 48 and 43, with 58 and 31 just visible in the station area. Continuing to the right but not visible were 55, 14, 17, 30, 15, 29, 28, 45, 13, 62 and 41.

Left: One of the spectacular mercury arc rectifiers at the Laxey sub-station. These vast glowing glass bulbs make a fascinating sight. Each unit has a cooling fan which can be seen at the bottom of the chamber.

When the MER first served Laxey in 1894 the terminus was just past today's sub-station on the south side of the valley. For the Ramsey extension an impressive viaduct was built across Glen Roy, linking the two sides of the valley. A well loaded car 21 and equally full trailer 48 cross the viaduct on their approach to Laxey station while working the 1.45pm service from Derby Castle to Ramsey on 26 June 1955.

Hugh Ballantyne

in the 1953 season. Dhoon and Ballaglass Glens were sold for £5660.

The winter service was becoming less and less viable with the increased competition from buses. This prompted the MER to promote a bill in 1953 to allow the withdrawal of the winter service. The bill, which was not debated until the summer of 1954, was defeated on the grounds that the tramway provided an important link for the residents of the Maughold area, which was not served by the buses. A proposal at the same time to allow fares to be increased was sanctioned.

The Island's tourist trade continued to fall away and, interestingly, so did the population, dropping from 55,253 in 1951 to 48,150 in 1961.

The Government commissioned a report on the future of its tourist business and one of the principal recommendations was the retention of both the MER and Snaefell lines, with assistance from the Government if required.

The renewal of the timbers on the Ballure viaduct meant single line working between Lewaigue and Ramsey. A collision on the singe line resulted in four passengers being treated in hospital but with only minor injuries.

The Snaefell line celebrated its 60th anniversary in August 1955, resulting in some very useful press coverage for both lines and indeed the Island as a whole.

The continuing fall in passenger numbers and resulting financial pressures by December of 1955 led the MER directors to tell the Government that they would be unable to operate the lines after the end of the following season and they offered to sell the

entire undertaking for £70,000.

A Government committee was appointed to look into the future of the line. They in turn appointed three specialists from the London Midland Region of British Railways, who came up with a figure of £674,000 to replace all the track, rolling stock and ancillary items and concluded that it would be cheaper to replace the service with double-deck buses.

The Light Railway Transport League organised what turned out to be a very well attended weekend convention in Laxey over Whitsun 1956. This generated a considerable amount of publicity for the line.

The specialist report was shelved in

Left: As part of the 1993 centenary celebrations, steam hauled specials were operated between Laxey and Dhoon Quarry. The two Milnes closed trailers 57 and 58 were hauled by steam railway No.4 *Loch*. Overnight storage was in the Laxey car sheds and the trio is seen here crossing Glen Roy viaduct and arriving in the station for the first train of the day. No.4's older sister No.2 *Derby* was hired by the contractor during the construction of the Laxey to Ramsey extension.

Bottom left: An 1894 power car stands at Laxey coupled to trailer 27, thought to be the present day 60. This trailer was damaged in the Laxey fire in 1930 but was repairable. The huge refreshment room, to the left of the picture, measured 140 x 40ft and was destroyed by fire on 24 September 1917.

Author's collection

Right: **Power car 27, earning its keep with the poles and wires department, shows off the temporary windscreens fitted to protect its driver from the elements. Normally housed at Laxey along with the two wiring trolleys that it has in tow, it is seen here at Laxey station pausing while its crew discuss their days work.**

Opposite page bottom right: **A picture that has been published before but is worthy of inclusion here. Two Bonner wagons are seen at Laxey with a 14-18 series power car. The rail chassis, which would have been about 16'0" long including couplings, are clearly visible with their 24" solid disc wheels. This picture is thought to have been taken on 14 August 1899 during a test run from Derby Castle. The IOMT&EP Co paid £536-11-2 for the three wagons which presumably included the rail chassis, of which there may only ever have been two.**

Isle of Man Railways

favour of a further review which was to investigate the possibility of retaining the line and operating at reasonable cost.

Meanwhile the MER petitioned Tynwald in July 1956 to accept an abandonment bill for the line but thankfully this was declined by a small majority. Instead Tynwald approved an indemnity bill against losses if the company continued to operate the line for the period up to 30 September 1957. This they agreed to do.

In excess of a quarter of a million people used the line in 1956, the operating loss totalling a mere £3,169.

A third group of experts inspected the line during July 1956 and reported that the cost of keeping the line open was far lower than the previous figure.

Finally, in August 1956, a Tynwald committee and directors of the MER met to discuss the future and an offer by Tynwald to buy the line for £50,000 was accepted by the MER. The committee reported back to Tynwald in November 1956 in favour of purchase and the implementation of a ten-year renewal plan for the Douglas to Laxey and Snaefell lines.

The proposal was debated in the House of Keys on 12 December 1956, the main objections coming from a member of the IOMR/Isle of Man Road Services board who reckoned that ten buses were all that was needed to replace the line. In support of retention was the fact that seventy percent of visitors to the Island travelled on the line and the positive effect on Ramsey that this generated. The vote was seventeen to four in favour of purchase.

The Bill was formally signed on 17 April 1957 and the MER was saved. The Manx Electric Railway Board of Tynwald was set up in May 1957 to manage the line and on the morning of 1 June the Lieutenant Governor of the Island drove a decorated car 32 from Derby Castle to Groudle to commemorate the transfer to Government ownership.

The intention of the Board was to carry out renewal of the Douglas-Laxey section in the first seven years followed in the next three by the

The palm trees of Laxey station greet car 6 arriving from Ramsey in 1981. The Mines Tavern is seen in the background; this was the former Station Hotel which was sold by the MER in 1957. Originally it was the Mine Captain's house and had stables behind for the horses used on the Glen Road Horse Freight Tramway (part of the great Laxey mines complex). The goods shed built in 1903 is also visible.

Opposite page: The 1993 'Year of Railways' brought enthusiasts from all over the world flooding to the Island. It had been many years since the author had witnessed scenes such as this, as passengers struggle to find seats on trailer 42 and one of the winter saloons on 5 September 1993. The dog seems content to wait for the next car which in true MER style will doubtless be just a few minutes behind.

The recently restored Laxey station building. The conservatory extension to the booking office was revealed by the removal of encroaching vegetation. The inside was fully restored and historical views of the MER now hang on the walls.

Snaefell line. Later in June 1957 200 tons of rail and associated sleepers were ordered, these arriving at Ramsey during October of the same year.

Earlier, in February 1957 the MER had been made aware of two former Llandudno and Colwyn Bay Electric Railway cars that were about to be broken up at Colwyn Bay. They were double-deck cars and were deemed unsuitable because of the considerable amount of re-building required. The height of the MER overhead wires prevented their use as double-deckers.

In order to balance the tourist employment during the summer, Tynwald operated winter employment schemes and these were used to good effect on the MER, the repair of fences and drainage and weeding of the line all being carried out by these groups. The Derby Castle depot yard and Laxey and Ramsey stations were tarmacked.

Passenger figures dropped to around the 200,000 mark in 1957, Tynwald approving £9000 to offset losses and a further £4000 for more trackwork repairs.

The Board decided that the cars should be painted in a corporate livery and agreed on green and white, the first repainted car appearing on 24 December 1957. New uniforms were issued to staff at about this time. Much has been written about the livery elsewhere; suffice to say that it was abandoned in 1959, with just fourteen cars being repainted.

Following a row about the finances among the Board members, a motion to close the Ramsey line completely was defeated by just one vote in May 1958. It was agreed, however, to run the entire line purely as a tourist attraction, restricting the service to between May and September. A new eleven return journey 10.00 to 18.00 service was introduced in June and uproar ensued.

The Chairman and three Board members resigned on 18 June and Douglas Corporation and Isle of Man Road Services sought licences to operate bus services to Groudle and other areas served by the MER, on the grounds that the tramway service was inadequate.

The regrouped Board was formed on 8 July 1958 and announced that the full twenty return journey timetable would be reinstated all year round. This enabled the Post Office contract to be renegotiated with improved rates. The new service was introduced from 12 July.

The Snaefell line benefitted from several improvements including the refurbishment of the Summit Hotel.

Souvenirs became an important source of revenue, an excellant range of colour postcards being among the items introduced.

Two hundred tons of new rail arrived in Douglas from the mainland in

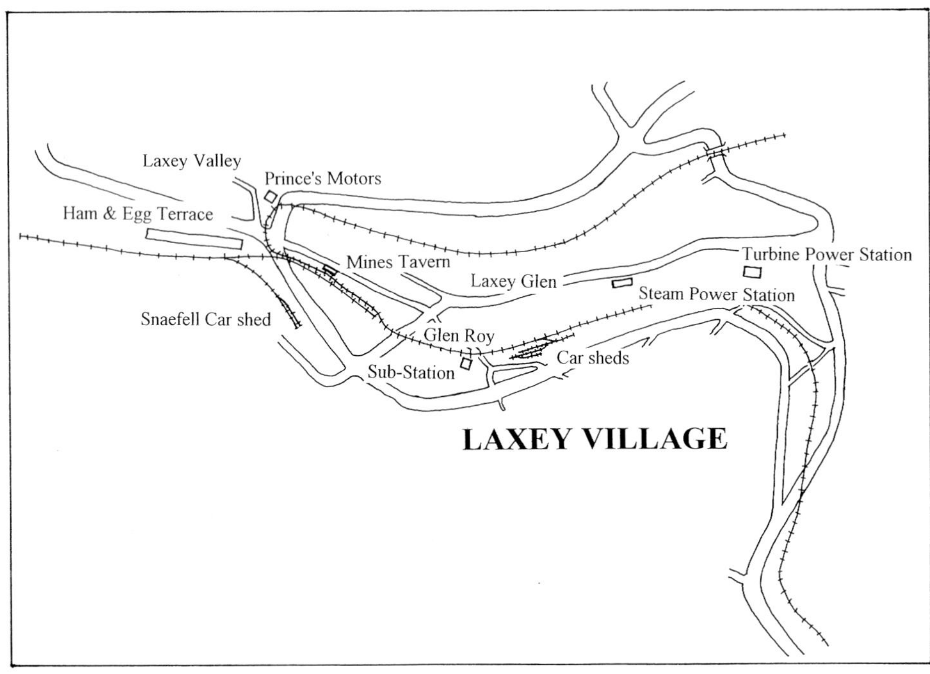

Bottom left: **During the summer months an evening service operates on the line. The last car leaves Ramsey at 21.15, Laxey at 22.00 and arrives at Derby Castle at 22.30. On Friday 22 August 1997 car 22 forms this working and was captured on film at Laxey, after express permission from the motorman to make use of a flashgun.**

Bottom right: **1995 marked the centenary of the Snaefell Mountain Railway and a special stamp issue was launched on 8 February, the first day covers travelling by various modes of transport from the Post Office in Douglas to the summit of Snaefell. Car 22 was used for the Derby Castle leg of the journey and is seen here at Laxey with mail van No.4 in tow. Snaefell car 2 which was also used is seen standing in front of its temporary home, while the depot was rebuilt.**

September and a further one hundred tons early the following year. Some six hundred tons had arrived by 1962 with five hundred tons in place by 1964.

Forty gallons of paint were used to repaint the Ballure viaduct in 1959. An inspector from the English Department of Transport gave the line a clean bill of health after carrying out a detailed inspection of the trackwork.

Shortly after nationalisation the MER board invited tenders from Thompson Houston & Co. to build four new tramcars. The invitation was passed to The Clayton Equipment Co. who submitted their tender on 9 July 1959. The cars would have had either 4 x 30hp motors at 550 volts each and wired in parallel, giving a maximum speed of 25mph or 4 x 60hp motors at 275 volts each wired in two series pairs and giving a top speed of 35mph. The price would have been £17,028 each, interestingly with optional extras quoted as dynamic braking £150, deadman's feature £20, packing and shipment £420. The cars were never built.

During the late 1950's and early 1960's thoughts were turned to the advertising and general publicity of the line. Cinema advertising was introduced, Douglas Corporation buses carried posters and the Manx Radio weather forecasts were sponsored.

The results of these initiatives were very positive, the Snaefell line benefitting in particular with its passenger figures rising from 68,000 in 1958 to 124,000 in 1961, dropping

back by 13,000 in 1962. Unfortunately, the weekly steamer from Fleetwood that, over the years, had brought thousands of day trippers to the Island and to the tramway, ceased operating in September 1961.

To improve connections at Derby Castle the horse tramway and promenade bus services timed their departures to coincide with MER arrivals. Other improvements included repainting all the overhead wire poles in dark green and treating the lineside buildings and fencing to a facelift. The whole line was given a general tidy up and £180,000 had been spent by the end of 1962.

Following final elimination of the green and white livery in 1963 all senior staff were provided with new uniforms,

In August 1895 the Snaefell Mountain Railway opened between Laxey and near the summit of the Island's highest mountain. The line was purchased by the IOMT&EP Co. in December 1895. Seen here at the Summit Hotel are passenger car 1, works/coal car 7 *Maria* and the Hurst Nelson 4ton open wagon.

Manx National Heritage

replacing those issued a few years earlier. Financial restraints put paid to plans to rebuild the Laxey station area and to build a chairlift over Douglas Harbour.

Fare increases in 1964 were perhaps more realistic than on previous occasions as visitors were often able to gain discounts on the advertised fares with vouchers from their hotels.

Top left: **The IOMT&EP Co. hired Isle of Man Railway No.2 *Derby* and Manx Northern No.1 *Ramsey* for the construction of the Laxey to Ramsey line. Here we see construction workers gathered round No. 2 during a break in activity, somewhere between Laxey and Minorca. The IOMR charged £2-10-0 per day for the locomotive, plus wages for the driver, fireman and a cleaner. Similar charges were made by the MNR.**

Isle of Man Railways

Top right: **For the construction of the Laxey to Ramsey line the contractor, Morrison and Mason, purchased an 0-4-0 saddle tank locomotive from Andrew Barclay of Kilmarnock (works No. 713 of 1892). On completion of the work the locomotive was sold to Douglas Corporation for use on the Baldwin Reservoir line, where it received the name *Injebreck*. It is seen here at the dam face on the reservoir line .**

Manx National Heritage

Opposite page bottom: **Car 22 and trailer approach Ballaragh with a Derby Castle to Ramsey service in August 1997. At this point the pair are climbing at 1:24 towards Bulgham summit.**

Right: **Car 14 with trailer 52 stands just short of the Bulgham earthslip site. The wooden shelter and steps up to the road are visible behind the tram. The photograph is dated July 1967 and must therefore have been taken a few days before through services resumed.**

A D Packer

The relaying of the Douglas to Laxey section, apart from at least one short stretch between South Cape and Laxey car sheds, was completed in 1965. Some attention had also been paid to the Ramsey line, mainly re-sleepering north of Dhoon.

The station and depot at Derby Castle came under threat when Douglas Corporation purchased the entertainment complex for redevelopment. The Corporation apparently did not want horse-drawn or antiquated electric trams in the vicinity of their state of the art complex. Onchan Head was suggested as the site for a replacement depot.

Thankfully, when the final plans for the complex were revealed, the MER was shown to be staying put, the new building spreading over the tracks on its upper levels with a bridge to the sea side of the road. The site was cleared in 1966 and construction of the new complex started in 1967.

The new MER Board had continued to operate the freight service, part of which was a collection and delivery link with all areas of Douglas and around the major stopping points on the line. By 1966 the lorries used for this service needed replacing and, as it was considered uneconomical to do so, the service ceased.

Tynwald set up a committee to look at transport to, from and on the Island. The 125 page report was published in May 1966 and while it dealt mainly with sea and air travel, the MER featured sufficiently for a recommendation that the Laxey to Ramsey line should be closed. This recommendation was not accepted by Tynwald, the likely reduction in passengers on the Douglas-Laxey section being among the considerations.

A seamen's strike in 1966 reduced the number of visitors to the Island and hence the passenger figures for the MER. Interestingly the winter service gained extra revenue when losses by the bus company led to a reduction in their Ramsey service.

On 20 January 1967 part of the retaining wall at Bulgham, where the line runs at 600ft above sea level, collapsed and a further collapse on 28 January took away the southbound track and forced this section of the line to close.

The MER Board acted quickly and within a few days of the first collapse, steps were placed in the wall separating the tramway from the road, crossovers to allow the trams to reverse either side of the breach were installed and services were resumed with passengers walking the short

Ten years before the Bulgham earthslip a 4-9 series car hauls trailer 41 along the ledge between the road and sea. There was clearly a problem on the northbound track as this pair are working 'wrong line'.

R J S Wiseman

Opposite page: **After leaving Laxey the line climbs steadily at gradients up to 1:24 for some two and a half miles before reaching Bulgham summit, where the line turns inland to follow the southern slope of Dhoon Glen towards the station. Carefully negotiating the sharp curve past the summit point is car 21 heading for Ramsey during May 1997.**

Car 20 and trailer rumble past Bulgham, along the ledge that has on more than one occasion caused concern. To the left of the picture it is clear just how close the line comes to the edge and it is hard to imagine that there is a road between the tram and the rock face to the right of the view. The road was closed for a few days during the recent winter after a dry stone wall collapsed onto it from above.

distance from one tram to the other. This increased overall journey times by five minutes.

Thankfully but perhaps suprisingly the Government announced that full repairs would be carried out to the trackbed and the contractors moved onto the site on 1 May 1967. Work continued apace and at 07.00 on 10 July, the first through car left Derby Castle and, with the southbound corresponding working leaving Ramsey at 07.15, the MER was back in full swing.

General repairs to the track and equipment continued, replacement poles still coming from the Douglas Head tramway. For the record there are 904 poles on the line. Some former IOMR rail was also obtained following the closure of the Ramsey and Peel lines in 1968.

The Light Railway Transport League assisted in the search for secondhand tramcars for the MER, particularly between early 1970 and the end of 1973. 950mm gauge cars from Caghari, Sardinia which were on Brill 77 bogies were considered in January 1970. Later that year, trolleybuses were replacing trams in the Spanish city of Valencia but, by the time consideration was given to the displaced trams, they had already been disposed of because the trolleybuses were using the old tramway depot.

In November 1970, eight power cars and nine trailers from the metre-gauge Brussels tramway were considered. Interestingly, the LRTL suggested that, as these cars represented good value

with a total cost of only £8,700, the MER should be re-gauged to accommodate them. It was also noted that further cars were likely to be available in the future.

Complete trams from Geneva and bogies only from Nagoya in Japan were among 1971's offerings. The 3'0" gauge Majorca tramways were approached in November 1973 but, far from having trams for sale, were also trying to buy and were not finding anything suitable. In the same month 2'6" gauge cars from Ostrava, Czechoslovakia were offered; these whilst needing re-gauging were the right voltage.

In the midst of all this the MER

considered motorising the two closed trailers 57 and 58, with facilities for one man operation. Two open power cars would probably have been scrapped and their bogies used, although 57 and 58 ride on Brill bogies which are probably the best on the line; motorising these did not appear to have been considered. This, like all the other options, was not carried forward.

On the evening of 2 August 1973 fire broke out in the Derby Castle leisure complex; fifty people died and scores were injured. *Mann Tram* for August/September 1973 reports that MER staff played hoses onto the sheds and tramcars to save them from destruction. While the area was made safe trams terminated at Port Jack.

Top left: **Dhoon Glen station is probably the busiest on the Laxey-Ramsey section. The impressive Dhoon Glen Hotel once stood behind the photographer but was destroyed by fire in April 1932. However, weary walkers returning to the station from the depths of the glen will find a welcome cup of tea at 'Jeans Place'. The station shelter was demolished in 1985 and replaced by a wooden version in 1987.**

Top right: **Winter saloon 20 hauls trailer 43 past open car 29 at Dhoon Quarry in August 1965. The remains of the quarry sidings can be seen to the right of the picture. The cottage known as 'Creosote Cottage' was demolished about 1979/80.**

Author's collection

Opposite Page Bottom: **Trailer 59 and car 2 at Dhoon Quarry on 20 May 1956. To the left of the car are two unidentified dreadnought stone wagons and three 6ton opens, the third of which is number 7. Visible above number 7 are two Milnes plate bogies sitting at the end of the siding reached via the points just to the right of 59. These sidings were once part of the busy rail terminal serving two quarries, one each side of the line. To the left a narrow gauge line worked into the IOMT&EP Co quarry while on the right an aerial ropeway served that operated by the Highways board.**

G B Blacklock

Right: **The MER timetable dated 21 September 1968. It is interesting to note that the full journey was given 70 minutes compared with 75 minutes today. The fares are worth a look; imagine being able to travel to and from Ramsey for just 4 shillings (20p)!**

MANX ELECTRIC RAILWAY — Douglas, Laxey and Ramsey

COMMENCING SATURDAY, 21st SEPTEMBER, 1968, and until further notice

WEEKDAYS — Sunday Service Suspended

Connecting Bus from Central Bus Station, Lord Street		8.35	9.35	11.35	1.05	2.35	4.05	5.35		8.35
DOUGLAS (Derby Castle)	dep.	7.00 a.m.	8.45	9.45	11.45	1.15 p.m.	2.45	4.15	5.45	7.15	8.45
Majestic Hotel	,,										
Groudle Glen	,,										
Baldrine	,,										
Garwick Glen	,,										
LAXEY	,,	7.25	9.10	10.10	12.10	1.40 p.m.	3.10	4.40	6.10	7.40	9.10
Dhoon	,,										
Glen Mona, B'glass	,,										
RAMSEY (Plaza)	arr.	8.10	9.55	10.55	12.55	2.25	3.55	5.25	6.55	8.25	9.55
RAMSEY (Plaza)	dep.	7.15	8.25	10.25	11.50	1.25	2.50	4.25	5.45	7.20	8.25
LAXEY	,,	8.00	9.05	11.05	12.35	2.05	3.30	5.05	6.30	8.00	9.05
DOUGLAS (Derby Castle)	arr.	8.30	9.35	11.35	1.05	2.35	4.00	5.35	7.00	8.30	9.35
Connecting Bus to Victoria Pier and Cent'l Bus Station Lord Street			8.36	9.36	11.36	1.10	2.36	4.06	5.40	7.21	

Return Fares

RAMSEY to :		DOUGLAS to :	
Ballajora	9d.	Groudle	1/-
Ballaglass	1/3	Garwick	1/3
Glen Mona	1/6	LAXEY	2/-
Dhoon	1/9	Dhoon	2/6
LAXEY	2/3	Glen Mona	2/9
Fairy Cottage	2/6	Ballaglass	3/-
Halfway H'se	3/-	Ballajora	3/6
DOUGLAS	4/-	RAMSEY	4/-

WEEKLY TICKETS and CONTRACT TICKETS available at Reduced Rates

Goods, Merchandise and Parcels conveyed between all Stations

General Offices : 1 Strathallan Crescent, Douglas
Tels. Douglas 4549 ; Laxey 226 ; Ramsey 2249

H. GILMORE, General Manager

The Ramsey line was again under threat when on 18 March 1975 Tynwald debated its future. The vote went in favour of the entire system closing for the winter on 1 October 1975, the Ramsey line for good.

The protest was vigorous and caused the Government to make a press statement on 4 July that the winter closure would go ahead but that the Ramsey line would not be physically abandoned. At a further meeting on 8 July the overall proposal was amended to read just that the line would cease all operations between 1 October 1975 and 30 April 1976.

At 16.25 on 30 September car 19 hauling trailer 57 left Ramsey for Derby Castle with what could so easily have been the last tram from the town. Buses took over the school services.

The need to get leaflets printed for the following season meant that a decision had to be made about the Ramsey line and on 10 December 1975 Tynwald announced that it would remain closed during 1976. Interestingly car 7 remained at Ramsey and made daily outings to Laxey until March 1976, when the press caught on that it was being used for staff transport!

1976 was a general election year on the Island, the campaign being dominated by the future of the railways. Traffic on the MER was down by twenty seven percent in the year. The new Government met for the first

time on 23 November, deferring any decision until January and eventually debating the MER on 16 February.

The vote this time was overwhelmingly in favour of reopening the line as soon as possible. A contractor was engaged to carry out any necessary repairs.

On Saturday 25 June 1977 a short ceremony was held in Laxey station before car 20 also with trailer 57 departed for Ramsey. Two other pairs of cars took invited guests and the public service started at 15.15.

The Manx Electric Railway Board took over the steam railway on 13 January 1978 and became Isle of Man Railways. Much of the rolling stock carried this title, often on specially added boards. Rover tickets were introduced allowing freedom of the

Dhoon Quarry in May 1997 with winter saloon 21 passing the permanent way flat 21 and 6ton open number 8. Much of the quarry evidence has now disappeared, the permanent way siding and a loop for reversing steam operated trains from Laxey being all that remain.

Opposite page: Car 20 approaches the Ballagorry overbridge with a southbound service during May 1997. The bridge, from where this picture was taken, now houses one of the lines electricity sub-stations, replacing the one at Ballaglass. The latter was housed in the former power station which has now been converted into a private residence.

The 1992 built replacement car 22 passes Dhoon Quarry with a Ramsey bound service in May 1997. The station area is to the left of the picture and the pile of sleepers form the base of the water tower used during steam operations from Laxey. The tank itself had been on loan to Ballasalla while the home tank was repaired.

entire electric and steam systems.

Inclement weather during the 1978 season meant that more passengers headed for the enclosed carriages of the steam railway than the open trailers on the electric. This however did not deter the Government which continued with repairs on both systems to the tune of around £850,000. Attempts were made at the end of 1978 to sell off four power cars and trailers but no serious buyers came forward and the idea was abandoned.

The Isle of Man Railway Society sponsored the opening of an Electric Railway Museum in Ramsey in May 1979. The weather was better which meant a good season and on 17 October Tynwald agreed to retain the Ramsey line for at least the next five years.

Isle of Man Railways and Isle of Man Road Services joined forces under the banner of Isle of Man National Transport in March 1980. The new organisation immediately announced the reintroduction of the winter service on the Manx Electric from the 1980/81 winter. This followed the renewal of the trackwork through Laxey station.

In March 1980 Douglas Corporation demolished the impressive canopy over the horse tramway terminus, a cost cutting exercise that prompted much critiscism.

Following the 1981 general election on the Island, the new Government recognised the importance of the railways as a tourist attraction, although visitor numbers to the Island continued to fall.

Land erosion problems to the south of Ballure again underlined the commitment of the Government. £100,000 was spent on repairs carried out by the Highways Board; the adjacent road was considered to be equally at risk of collapse.

Tynwald passed a Passenger Transport Bill in early 1982, the new title, Isle of Man Passenger Transport Board, becoming legally effective in April 1983. Car 6, which happened to be in the paint shops at the time, emerged displaying 'Isle of Man Passenger Transport'.

The 1986 general election saw a change in the format of ministerial positions, the railways and buses becoming part of the Department of Tourism and Transport. Mr William Jackson, who had taken over as Chief Executive in 1977, announced his retirement from the post, now designated Chief Officer of the

Top: **Approaching Ballaglass from the north the line goes through a reverse curve. Car 21 with its trailer just visible eases round the curves towards the station. The walk through the nearby glen offers peace and quiet and the opportunity to look at the 70ft diameter bridge span that carries the MER over the river.**

Bottom left: **Car 20 and trailer are approaching Dreemskerry station past the site of the former siding that served the now disused Ballajora quarry, the crossover for which can be seen behind the trams.**

Below: **The certificate of incorporation of the Manx Electric Railway Company Limited dated 12 November 1902. The original of this survives in the railway collection.**

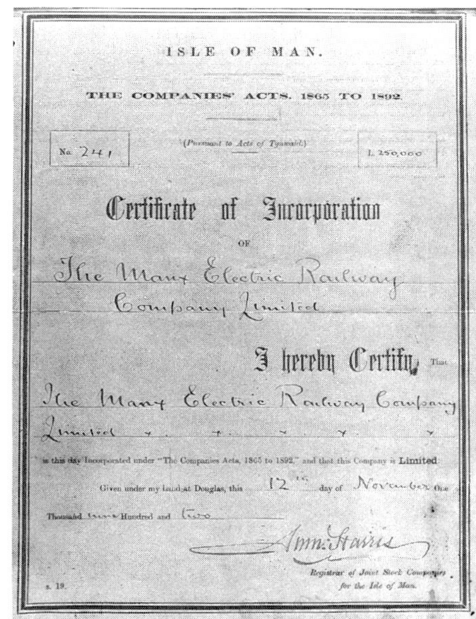

Winter saloon 21 complete with trailer approaches Bellevue with a Ramsey to Derby Castle service in May 1997. To the right of the car can be seen the Queens Pier in Ramsey and parts of the harbour walls. The tram is just a few minutes into its 75 minute journey.

Department of Tourism and Transport. Mr Robert Smith was appointed Transport Executive, taking up the appointment on 1 December 1987.

Mr Smith immediately set about introducing a fully integrated transport system on the Island, with the buses, the Manx Electric and Snaefell lines and the Douglas to Port Erin steam railway, which started running daily, all coming together as one system. The result was impressive with seventeen percent more passengers using the MER in 1988 than in the previous year, leading to improvements in the service for 1989 both during the day and the evening.

Laxey station was highly commended in the 'best restored station' category of the 'public and commercial' section of the 1988 Ian Allan, British Railways and Railway Heritage Trust awards.

New purpose-built staff amenities were provided at Derby Castle depot between the upper and lower car sheds.

The computerisation of the ticketing systems on British Railways led to the sale of the machines that produced the traditional Edmondson card tickets that had been supplied to both the MER and IOMR for many years. Fortunately a suitable replacement supplier was soon found.

The electric sub-station at Ballaglass was replaced by a modern installation, housed under the bridge at Ballagorry, which started providing full service in early 1990. The Ballaglass building has since been converted to residential use.

The Government reinforced its commitment to the line as a tourist attraction in late 1990. Car 22 had been completely refurbished in the run up to the season. On the evening of 30 September it returned to the depot and was parked in the lower shed in the company of sister car 19 and toastrack 14. An overheated resistor on the car caught fire, completely destroying the body apart from the Douglas end third. Within a few months it was announced that the car would be rebuilt using the original chassis. McArds of Port Erin completed all the carpentry while the railway staff rebuilt the electrical parts. The 'phoenix' car 22 entered service on 13 May 1992.

The number 1 shed at Derby Castle received a new roof during 1990 and the same roof received further attention the following winter after the fire that destroyed car 22. Many similar projects have taken place since, on all the railways of the Island.

The Plaza Cinema (formerly Palace Concert Hall) adjacent to Ramsey station and once owned by the MER was demolished in the early months of 1991, the site becoming a car park.

Left: **Photographed sometime between 1899 and 1914 a 19-22 series power car hauls a trailer over Ballure Viaduct. This view is now obscured by the trees of the Ballure plantation. Queens Pier and the north east coast of the Island are visible in the distance.**

Author's collection

Bottom left: **Francis Morton & Co, of Garston, Liverpool, manufactured this viaduct in 1899, bridging the final gap and allowing completion of the line through to the centre of Ramsey. Before this the line terminated at the far end of the viaduct and the existing Ramsey car shed was originally sited to the left of the picture. Car 20 with trailer in tow observes the speed restriction as it crosses on its way into the town.**

Right: **The winter saloons provide the backbone of the Ramsey service all year round. Visitors to the Island during early and late season will often find these cars working solo. Leaving Ramsey past the backs of the houses in Waterloo Road is car 19 with the 10.00 to Derby Castle.**

Opposite page bottom right: **The Ramsey Pier Tramway opened in 1899 with hand propelled passenger cars and over the years saw various replacements including several petrol driven vehicles. The line closed in September 1981 and the stock was stored for a while before being moved to the MER museum adjacent to Ramsey station for safekeeping and restoration. This locomotive built by F C Hibberd & Co Ltd in 1937 at their Park Royal works in London is a Y type 'Planet' 4-wheeled petrol driven machine (works No 2027) and is seen here with its modified 'Steam look' bodywork. The passenger car came from the same builder. The pair are seen on MER track while working a special from Ramsey station to Walpole Road in May 1989.**

The 1993 centenary of the Manx Electric was fast approaching and with this in mind Mr Alan Corlett was appointed centenary coordinator in 1991.

In recognition of the vast programme of special events, 1993 was designated 'Year of Railways'. The special events began with a launch on Easter Saturday and ran through to the end of October.

Car 9 became the Island's first illuminated tram at the beginning of the season. Some 1600 individual bulbs were fitted to the outside, along with special roof-mounted illuminated panels pronouncing '1893 Manx Electric Railway Centenary 1993'. These boards have since been changed a number of times, for the Snaefell centenary, Groudle centenary and currently carry 'Steam 125' in preparation for the 1998 season.

Steam operated on the MER between Laxey and Dhoon, where a special siding and watering facilities had been installed. Motive power was usually the steam railway's number 4 *Loch*.

The former goods shed at Ramsey was converted into a visitor centre during the year, trailer 59 forming part of the display.

Tuesday 7 September was centenary

Left: **Although the main depot for the MER is at Derby Castle, sheds for overnight accommodation were provided at Laxey and Ramsey. The Laxey shed still sees regular use but that at Ramsey is in need of restoration. Deep inside Ramsey shed we see car 6 awaiting its next turn of duty. These cars lost their twin windscreens in the late 1960's in order to improve driver visability.**

Hugh Ballantyne

Bottom left: **Near journey's end, car 19 with trailer glides into Ramsey past the former car shed, which at this time was in use as a museum. The track layout for the car shed is interesting as it involves a double shunt move to reach the left hand track.**

Opposite page bottom right: **One of the many items preserved in the Ramsey visitor centre is this painted advertisment, presumably from somewhere in Douglas or Derby Castle itself.**

day and, following a short ceremony at Derby Castle, car No.1 was driven to Groudle by the Governor, Air Marshal Sir Lawrence Jones under the watchful eye of engineering superintendant Mr Maurice Faragher. At Groudle a centenary plaque was unveiled and refreshments were served to the invited guests in the adjacent hotel.

The enormous success of these events put the Isle of Man Railways firmly back on the map, the railway press giving huge amounts of space to detailed coverage.

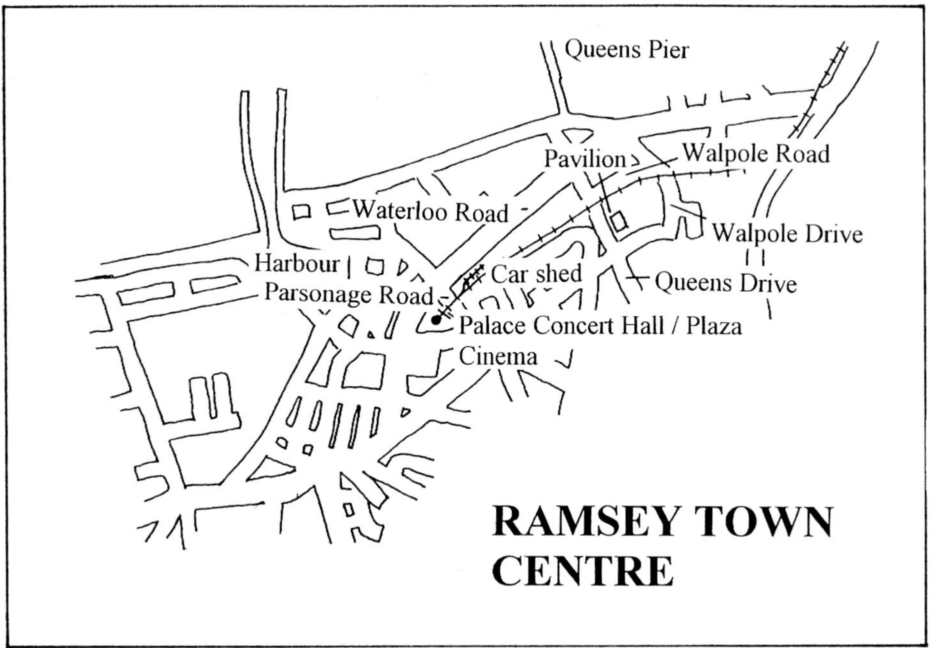

RAMSEY TOWN CENTRE

Considerable lengths of track were relayed during the 1993/94 winter; much of this relaying made use of new 75lb rail and represented the most significant track project for around 20 years.

Laxey station building was refurbished during 1994, the most notable change being the removal of thick vegetation to the right hand side revealing the conservatory extension to the ticket office.

The winter 1994/95 saw the replacement of the original Snaefell line car shed. A major refurbishment of the Summit Hotel was also carried out over the winter and early summer.

1995 was the centenary of the Snaefell line and an International Railway Festival was held on the Island to commemorate this event. MER trailer 56 had been converted to carry disabled passengers and was officially launched at the beginning of the season.

The possible purchase of one or more former Lisbon tramcars to provide spares for the MER fleet began to be investigated.

In December 1995 parts of the rock face adjacent to the road at Bulgham fell away, causing some concern but no damage to the MER trackwork. The road was closed and around two hundred tons of rock had either fallen or was deliberately dislodged to make the area safe.

The crossings at Halfway and outside Prince's Motors in Laxey were completely relaid during the 1995/96 winter. Bad weather in the north of the Island reinforced the value of the MER; many roads were impassable but the trams continued to operate providing a vital link with otherwise stranded residents.

Former Lisbon tramcar 360 arrived at Derby Castle depot on 3 June 1996, the first new power car since numbers 32 and 33 arrived in 1906; coincidentally 360 also dates from that year.

It had been announced in early 1996 that the top car shed at Derby Castle would be replaced over the coming winter at an estimated cost of

Left: **Three of the line's 4-wheeled vans stand at Ramsey. 3 arrived from G F Milnes in 1894, 12 in 1898/99 and 16 in 1908. All survive today, although 12 has been converted to a tower wagon.**

Author's collection

Bottom left: **A close-up view of Van 12 in original form. Several of the vans were delivered with a platform at one or both ends but many of these have since been removed.**

Author's collection

Bottom right: **The MER employed this Austin van (WMN 434) for its parcels collection and delivery service. Painted in the 1957 green and white livery it had shaded lettering and hand painted pictures on its sides. It is seen here at Ramsey in the company of rail vans 11 and 14.**

The Late Reverend John Parker

An unidentified traction engine hauls a Bonner wagon along the quayside in Ramsey. The road wheels measured about 3'6" giving an overall height of the vehicle in road mode of about 6'6". This increased to 7'6" when in rail mode. This picture is undated and it could be that the wagons saw road service long after they were discarded by the MER.

Manx National Heritage

£600,000. After much detailed planning, work started on clearing all rolling stock and useable accessories from the shed in preparation for its demolition and replacement with a modern, secure and protected structure.

Storage of the tramcars presented a major problem, the solution found being to put as many as possible under cover at Laxey and Ramsey. The remainder were lined up at the viaduct in Laxey and sheeted over.

Work on the new shed continued through the 1997 season, the final trackwork taking place over the recent winter. Plans are in hand to replace the cladding on Laxey and Ramsey car sheds in the next few years.

1998 marks the centenary of the line as far as Ballure and 1999 that of reaching Ramsey town centre.

Towards the end of January 1998 the House of Keys was going to hear the member for Ramsey ask what long term plans there were for the MER. Describing the railway as 'absolutely unique and a very precious asset' he

said that the government must do everything it can to maintain it. He would be asking the Minister for Tourism and Leisure to assure him that adequate plans were being made for the line for, say, the next 25 years.

The Manx Electric and Snaefell lines, as well as the steam railway, can surely look forward to a more certain future than they have on occasions in the past. It would have been so easy for the entire system to have become history, with this remarkable line lost forever. Long may it survive.

POWER CAR FLEET LIST

Car No.	Built by	Year	Bogies	Motors	Car Type	No. of seats	Length	Width	Height	Notes
1	Milnes	1893	Milnes S3 [1]	2 x 25hp	Unvestibuled saloon	34	34'9" 10.59m	6'6" 1.98m	11'0" 3.35m	In service [2]
2	Milnes	1893	Milnes S3 [1]	2 x 25hp	Unvestibuled saloon	34	34'9" 10.59m	6'6" 1.98m	11'0" 3.35m	In service
3	Milnes	1893	Milnes S3 [1]	2 x 25hp	Unvestibuled saloon	34	34'9" 10.59m	6'6" 1.98m	11'0" 3.35m	Destroyed Laxey 1930
4	Milnes	1894	Milnes S3 [3]	2 x 25hp	Vestibuled saloon	36	34'8" 10.56m	6'3" 1.90m	11'0" 3.35m	Destroyed Laxey 1930
5	Milnes	1894	Milnes S3 [1]	2 x 25hp	Vestibuled saloon	32	34'8" 10.56m	6'3" 1.90m	11'0" 3.35m	In service
6	Milnes	1894	Milnes S3 [1]	2 x 25hp	Vestibuled saloon	36	34'8" 10.56m	6'3" 1.90m	11'0" 3.35m	In service
7	Milnes	1894	Milnes S3 [1]	2 x 25hp	Vestibuled Saloon	36	34'8" 10.56m	6'3" 1.90m	11'0" 3.35m	In service
8	Milnes	1894	Milnes S3 [1]	2 x 25hp	Vestibuled saloon	36	34'8" 10.56m	6'3" 1.90m	11'0" 3.35m	Destroyed Laxey 1930
9	Milnes	1894	Milnes S3 [1]	2 x 25hp	Vestibuled saloon	36	34'8" 10.56m	6'3" 1.90m	11'0" 3.35m	In service [4]
10	Milnes	1895	Milnes S3	2 x 25hp	Vestibuled saloon	46	35'7" 10.85m	6'9" 2.05m	10'0" 3.04m	To freight car 26 [5]
11	Milnes	1895	Milnes S3	2 x 25hp	Vestibuled saloon	46	35'7" 10.85m	6'9" 2.05m	10'0" 3.04m	To freight car 21 [5]
12	Milnes	1895	Milnes S3	2 x 25hp	Vestibuled saloon	46	35'7" 10.85m	6'9" 2.05m	10'0" 3.04m	To freight car 22 [5]
13	Milnes	1895	Milnes S3	2 x 25hp	Vestibuled saloon	46	35'7" 10.85m	6'9" 2.05m	10'0" 3.04m	To freight car 23 [5]
14	Milnes	1898	Milnes S3	4 x 20hp	Cross bench open	56	35'5" 10.79m	6'3" 1.90m	10'6" 3.20m	Stored
15	Milnes	1898	Milnes S3	4 x 20hp	Cross bench open	56	35'5" 10.79m	6'3" 1.90m	10'6" 3.20m	Stored

POWER CAR FLEET LIST (continued)

Car No.	Built by	Year	Bogies	Motors	Car Type	No. of seats	Length	Width	Height	Notes
16	Milnes	1898	Milnes S3 [6]	4 x 20hp	Cross bench open	56	35'5" 10.79m	6'3" 1.90m	10'6" 3.20m	In service
17	Milnes	1898	Milnes S3	4 x 20hp	Cross bench open	56	35'5" 10.79m	6'3" 1.90m	10'6" 3.20m	Stored
18	Milnes	1898	Milnes S3	4 x 20hp	Cross bench open	56	35'5" 10.79m	6'3" 1.90m	10'6" 3.20m	In service
19	Milnes	1899	Milnes S3 [7]	4 x 20hp	Winter saloon	48	37'6" 11.43m	7'4" 2.23m	11'0" 3.35m	In service
20	Milnes	1899	Milnes S3 [7]	4 x 20hp	Winter saloon	48	37'6" 11.43m	7'4" 2.23m	11'0" 3.35m	In service
21	Milnes	1899	Milnes S3 [7]	4 x 20hp	Winter saloon	48	37'6" 11.43m	7'4" 2.23m	11'0" 3.35m	In service
22	Milnes	1899	Milnes S3 [7]	4 x 20hp	Winter saloon	48	37'6" 11.43m	7'4" 2.23m	11'0" 3.35m	Burnt out 1990 [8]
22	McArds (Port Erin)	1992	Brill 27Cx	4 x 25hp	Winter saloon	48	37'6" 11.43m	7'4" 2.23m	11'0" 3.35m	Replacement [8]
23	IOMT&EP Co. Ltd	1900	(Milnes S3)	(4 x 20hp)	Locomotive	--	20'6" 6.24m	7'6" 2.28m	11'0" 3.35m	Damaged in derailment 1914 [9]
23	MER Co. Ltd	1926	(Brill 27Cx)	(4 x 27.5hp)	Locomotive	--	34'6" 10.51m	6'3" 1.90m	10'0" 3.04m	Rebuild from above Serviceable [9]
24	Milnes	1898	Brush Type D	[10]	Cross bench open	56	35'5" 10.79m	6'3" 1.90m	10'6" 3.20m	Destroyed Laxey 1930
25	Milnes	1898	Brush Type D	[10]	Cross bench open	56	35'5" 10.79m	6'3" 1.90m	10'6" 3.20m	In service
26	Milnes	1898	Brush Type D	[10]	Cross bench open	56	35'5" 10.79m	6'3" 1.90m	10'6" 3.20m	In service
27	Milnes	1898	Brush Type D	[10]	Cross bench open	56	35'5" 10.79m	6'3" 1.90m	10'6" 3.20m	In service
28	ER&TCW Ltd	1904	Brill 27Cx [11]	4 x 25hp	Cross bench open	56	35'0" 10.66m	6'3" 1.90m	10'6" 3.20m	Stored

POWER CAR FLEET LIST (continued)

Car No.	Built by	Year	Bogies	Motors	Car Type	No. of seats	Length	Width	Height	Notes
29	ER&TCW Ltd	1904	Brill 27Cx [11]	4 x 25hp	Cross bench open	56	35'0" 10.66m	6'3" 1.90m	10'6" 3.20m	Stored
30	ER&TCW Ltd	1904	Brill 27Cx [11]	4 x 25hp	Cross bench open	56	35'0" 10.66m	6'3" 1.90m	10'6" 3.20m	Stored
31	ER&TCW Ltd	1904	Brill 27Cx [11]	4 x 25hp	Cross bench open	56	35'0" 10.66m	6'3" 1.90m	10'6" 3.20m	Stored
32	UEC	1906	Brill 27Cx	4 x 27.5hp	Cross bench open	56	35'0" 10.66m	6'3" 1.90m	10'6" 3.20m	In service
33	UEC	1906	Brill 27Cx	4 x 27.5hp	Cross bench open	56	35'0" 10.66m	6'3" 1.90m	10'6" 3.20m	In service
360	John Stephenson & Co	1907/8	Brill 27G E1	4 x 25hp	Vestibuled saloon	40	39'8" 12.09m	7'10" 2.38m		Serviceable [12]

Notes:

All power cars except 360 have 23'8" bogie centres.

[1] Brush type D bogies and 4 x 25hp motors fitted in place of Milnes S3 in 1903

[2] Oldest electric tramcar in the world still running on its original line.

[3] Received Milnes S3 bogies and 4 x 20hp motors from car 16 in 1899.

[4] Became the Island's first illuminated tram in 1993 as part of the centenary celebrations.

[5] Nos 10-13 were withdrawn from passenger service between 1902 and 1903 (it is possible that one or two cars ran in 1904) and converted for freight use becoming 26,21,22 and 23 respectively.

[6] Received Milnes S3 bogies and 2 x 25hp motors from car No.4 in 1899; Brush type D bogies and 4 x 25hp motors fitted in 1903.

[7] Nos 19-22 received Brill 27Cx bogies and 4 x 25hp motors from car Nos 29, 28, 30 and 31 respectively in 1904.

[8] Original car No.22 destroyed by fire at Derby Castle depot during the night of 30 September 1990. New body built by McArds of Port Erin and electrical repairs carried out by staff at Derby Castle. Re-entered service in May 1992.

[9] The original No.23, which borrowed bogies from No.17 when required, was damaged in a derailment in 1914 but not withdrawn until 1922 . It was not broken up and in 1926 was rebuilt and now borrowed bogies from the more powerful No.33. Disused after 1944, stored at Laxey until superficially restored in 1978; restored to working order in 1983/4 and for 1993. Named *Dr R. Preston Hendry* at Derby Castle on 25 May 1992.

[10] Nos 24-27 entered service in 1898 as trailers Nos 40-43, fitted with 4 x 25hp motors by the MER in 1903.

[11] Nos 28-31 received Milnes S3 bogies and 4 x 20hp motors from car Nos 20, 19, 21 and 22 respectively in 1904.

[12] Purchased from Lisbon Tramways in mid-1996.

TRAILER CAR FLEET LIST

Car No.	Built by	Year	Bogies	Car Type	No. of seats	Length	Width	Height	Notes
13	Milnes	1893	Milnes S1	Cross bench open	44	28'9" 8.76m	6'3" 1.90m	9'2" 2.79m	In service [1]
34	Milnes	1894	Milnes S2	Cross bench open	44	29'0" 8.83m	6'1" 1.85m	9'0" 2.74m	Destroyed Laxey 1930
35	Milnes	1894	Milnes S2	Cross bench open	44	29'0" 8.83m	6'1" 1.85m	9'0" 2.74m	Destroyed Laxey 1930
36	Milnes	1894	Milnes S2	Cross bench open	44	29'0" 8.83m	6'1" 1.85m	9'0" 2.74m	Stored
37	Milnes	1894	Milnes S2	Cross bench open	44	29'0" 8.83m	6'1" 1.85m	9'0" 2.74m	In service
38	Milnes	1894	Milnes S2	Cross bench open	44	29'0" 8.83m	6'1" 1.85m	9'0" 2.74	Destroyed Laxey 1930
39	Milnes	1894	Milnes S2	Cross bench open	44	29'0" 8.83m	6'1" 1.85m	9'0" 2.74m	Destroyed Laxey 1930
40	Milnes	1903	Milnes S3	Cross bench open	44	28'6" 8.68m	6'5" 1.95m		Destroyed Laxey 1930
40	English Electric	1930	Milnes S1	Cross bench open	44	28'8" 8.73m	6'5" 1.95m	9'7" 2.92m	In service [2]
41	Milnes	1903	Milnes S3	Cross bench open	44	28'6" 8.68m	6'5" 1.95m		Destroyed Laxey 1930
41	English Electric	1930	Milnes S1	Cross bench open	44	28'8" 8.73m	6'5" 1.95m	9'7" 2.92m	In service [2]
42	Milnes	1903	Milnes S3	Cross bench open	44	28'6" 8.68m	6'5" 1.95m		In service
43	Milnes	1903	Milnes S3	Cross bench open	44	28'6" 8.68m	6'5" 1.95m		In service
44	Milnes	1903	Milnes S3	Cross bench open	44	28'6" 8.68m	6'5" 1.95m		Destroyed Laxey 1930
44	English Electric	1930	Milnes S1	Cross bench open	44	28'8" 8.73m	6'5" 1.95m	9'7" 2.92m	In service [2]

TRAILER CAR FLEET LIST (Continued)

Car No.	Built by	Year	Bogies	Car Type	No. of seats	Length	Width	Height	Notes
45	Milnes	1899	Milnes S2	Cross bench open	44	28'8" 8.73m	6'5" 1.95m	9'10" 3.0m	In service
46	Milnes	1899	Milnes S1	Cross bench open	44	28'8" 8.73m	6'5" 1.95m	9'10" 3.0m	In service
47	Milnes	1899	Milnes S1	Cross bench open	44	28'8" 8.73m	6'5" 1.95m	9'10" 3.0m	In service
48	Milnes	1899	Milnes S2	Cross bench open	44	28'8" 8.73m	6'5" 1.95m	9'10" 3.0m	In service
49	Milnes	1893	Milnes S1	Cross bench open	44	28'9" 8.76m	6'3" 1.90m	9'2" 2.79m	In service
50	Milnes	1893	Milnes S1	Cross bench open	44	28'9" 8.76m	6'3" 1.90m	9'2" 2.79m	Stored
52	Milnes	1893	Milnes S1	Cross bench open	44	28'9" 8.76m	6'3" 1.90m	9'2" 2.79m	Body removed 1947 [3]
53	Milnes	1893	Milnes S1	Cross bench open	44	28'9" 8.76m	6'3" 1.90m	9'2" 2.79m	Stored
54	Milnes	1893	Milnes S1	Cross bench open	44	28'9" 8.76m	6'3" 1.90m	9'2" 2.79m	Stored
55	ER&TCC	1904	Brill 27CxT	Cross bench open	44	29'4" 8.94m	6'5" 1.95m	9'4" 2.84m	In service
56	ER&TCC	1904	Brill 27CxT	Cross bench open	44	29'4" 8.94m	6'5" 1.95m	9'4" 2.84m	In service [4]
57	ER&TCC	1904	Brill 27CxT	Unvestibuled saloon	32	32'9" 9.98m	6'9" 2.05m	10'8" 3.25m	In service
58	ER&TCC	1904	Brill 27CxT	Unvestibuled saloon	32	32'9" 9.98m	6'9" 2.05m	10'8" 3.25m	In service
59	Milnes	1895	Milnes S2	Unvestibuled saloon	18	22'2" 6.75m	6'9" 2.05m	10'10" 3.30m	Serviceable
60	Milnes	1896	Milnes S1	Cross bench open	44	28'9" 8.76m	5'9" 1.75m	8'7" 2.62m	In service

TRAILER CAR FLEET LIST (Continued)

Car No.	Built by	Year	Bogies	Car Type	No. of seats	Length	Width	Height	Notes
61	UEC	1906	Brill 27CxT	Cross bench open	44	29'4" 8.94m	6'5" 1.95m	9'4" 2.84m	In service
62	UEC	1906	Brill 27CxT	Cross bench open	44	29'4" 8.94m	6'5" 1.95m	9'4" 2.84m	In service

Trailer car numbering.

Trailers are listed here in current numerical order. These are the numbers they have carried since 1906 when the final power cars were delivered. Before 1906 many of the trailers were re-numbered, some several times when it was necessary to release numbers for new power cars. As an example of this, the first three trailers started life as 11-13, then became 23-25, then 28-30 and finally 49-51 (51 reverted to 13 for the centenary celebrations in 1993 and still carries this number).

Original trailers 40-43, built in 1898, were fitted with motors in 1903 and renumbered 24-27. They are shown in the power car list.

Notes:

[1]   Restored to 1893 condition and re-numbered from 51 for centenary year.
[2]   Replacements for the trailers destroyed in the Laxey fire were ordered in 1930. Although seven were destroyed, insurance only provided for three trailers and a replacement shed.
[3]   Body removed in 1947, chassis still in use as a permanent way flat.
[4]   Rebuilt with wheelchair access and fittings to carry disabled passengers. Re-entered service in 1995.

FREIGHT STOCK LIST

Fleet No.	Built by	Year	Bogies	Type	Length	Width	Height	Notes
1	Milnes	1894	--	6-ton open				In service
2	Milnes	1894	--	6-ton open				Withdrawn by 1957
3	Milnes	1894	--	6-ton van (with platform)	16'5" 5.01m	6'6" 1.98m	9'5" 2.88m	In service
4	Milnes	1894	--	6-ton van (with platforms)	16'5" 5.01m	6'6" 1.98m	9'5" 2.88m	In service
5	Milnes	1896	--	6-ton open	13'3" 4.04m	5'10" 1.77m	4'11" 1.5m	Broken up between 1968 and 1992
6	Milnes	1897/8	--	6-ton open	13'3" 4.04m	5'10" 1.77m	4'11" 1.5m	Withdrawn by 1957
7	Milnes	1897/8	--	6-ton open	13'3" 4.04m	5'10" 1.77m	4'11" 1.5m	In service
8	Milnes	1897/8	--	6-ton open	13'3" 4.04m	5'10" 1.77m	4'11" 1.5m	In service [1]
9	Milnes	1897/8	--	6-ton open	13'3" 4.04m	5'10" 1.77m	4'11" 1.5m	Withdrawn by 1957
10	Milnes	1897/8	--	6-ton open	13'3" 4.04m	5'10" 1.77m	4'11" 1.5m	Broken up 1965
11	Milnes	1898/9	--	6-ton van(with platforms)	12'7" 3.83m	6'6" 1.98m	9'10" 3.0m	In service Now without platforms
12	Milnes	1898/9	--	6-ton van(with platforms)	12'7" 3.83m	6'6" 1.98m	9'10" 3.0m	In service [2] Now a tower van with ridge roof
13	Milnes	1903/4	--	5-ton van	10'10" 3.30m	5'10" 1.77m	8'8" 2.64m	In service [3]
14	Milnes	1903/4	--	5-ton van	10'10" 3.30m	5'10" 1.77m	8'8" 2.64m	In service
15	MER	1908	--	6-ton van	12'10" 3.92m	6'8" 2.04m	9'3" 2.83m	Destroyed 1944

FREIGHT STOCK LIST (Continued)

Fleet No.	Built by	Year	Bogies	Type	Length	Width	Height	Notes
16	MER	1908	--	6-ton van	12'10" 3.92m	6'8" 2.04m	9'3" 2.83m	In service
17	Milnes/ Voss	1912	--	6-ton open				Withdrawn [4]
18	Milnes/ Voss	1912	--	6-ton open				Withdrawn [4]
19	MER	1912		12-ton stone wagon				Condemned as unfit 1933
20	MER	1912		12-ton stone wagon				Condemned as unfit 1933
21	Milnes/ MER	1895	Milnes	Freight motor car	35'7" 10.85m	6'9" 2.05m	10'0" 3.04m	Broken up 1926 [5]
21	MER	1926	Milnes	12-ton stone wagon	32'3" 9.84m	6'5" 1.97m	3'4" 1.01m	In service [6]
22	Milnes/ MER	1895	Milnes	Motor cattle car	35'7" 10.85m	6'9" 2.05m	11'2" 3.41m	Demotored 1912 Broken up 1927 [7]
23	Milnes/ MER	1895	Milnes	Freight trailer	35'7" 10.85m	6'9" 2.05m	10'0" 3.04m	Renumbered 22 in March 1927 Broken up late 1950's. [8]
24 [9]	MER	1910/12		Sheep trailer	32'9" 9.98m	6'10" 2.08m		Bogies to new 24 in 1924 Presumed broken up.
24	MER	1924		12-ton stone wagon	32'2" 9.80m	6'8" 2.03m	5'2" 1.57m	Withdrawn by 1956
25 [9]	MER	1910/12		Sheep trailer	32'9" 9.98m	6'10" 2.08m		Bogies to new 25 in 1924 Presumed broken up.
25	MER	1924		12-ton stone wagon	32'2" 9.80m	6'8" 2.03m	5'2" 1.57m	Withdrawn by 1956
26	Milnes/ MER	1895		Bogie freight trailer	35'7" 10.85m	6'9" 2.05m	10'0" 3.04m	Preserved [10]
52	Milnes	1893	Milnes S1	Engineers flat	28'9" 8.76m	6'3" 1.90m		In service

FREIGHT STOCK LIST (Continued)

Fleet No.	Built by	Year	Bogies	Type	Length	Width	Height	Notes
--		1893?	--	Tower wagon				Destroyed Laxey 1930 [11]
--		1893/4	--	Tower wagon				Destroyed Laxey 1930 [11]
--		?	--	Tower wagon				Destroyed Laxey 1930 [11]
--	?	189 ?	?	Platelayers bogie wagon				[12]
--	?	189 ?	?	Platelayers bogie wagon				[12]
--	?	189 ?	?	Platelayers bogie wagon				[12]
--	?	189 ?	?	Platelayers bogie wagon				[12]
--	?	189 ?	?	Platelayers bogie wagon				[12]
--	?	189 ?	?	Platelayers bogie wagon				[12]
--	?	189 ?	?	Platelayers bogie wagon				[12]
--	?	189 ?	?	Platelayers bogie wagon				[12]
--	?	189 ?	?	Platelayers bogie wagon				[12]
--	?	189 ?	?	Platelayers bogie wagon				[12]
--	?	189 ?	?	Platelayers bogie wagon				[12]
--	?	189 ?	?	Platelayers bogie wagon				[12]
--	?	189 ?	?	Bogie wire wagon				[12]
--	?	189 ?	?	Bogie cable wagon				[12]
--	Bonner Wagon Co.	1899	--	Bonner rail/road wagon	16'0" 4.88m	6'0" 1.82m	7'6" 2.28m	[13]
--	Bonner Wagon Co.	1899	--	Bonner rail/road wagon	16'0" 4.88m	6'0" 1.82m	7'6" 2.28m	[13]
--	Bonner Wagon Co.	1899	--	Bonner rail/road wagon	16'0" 4.88m	6'0" 1.82m	7'6" 2.28m	[13]

FREIGHT STOCK LIST (Continued)

Fleet No.	Built by	Year	Bogies	Type	Length	Width	Height	Notes
(1)		1930?	--	Tower wagon				[14]
(2)		19 ?	--	Tower wagon				[14]
(3)		19 ?	--	Tower wagon				Dismantled since 1977 [14]

Notes:

[1]   Currently carrying a generator for which a replica van top has been built [13'3" (4.04m) long, 5'10" (1.77m) wide and 7'2" (2.19m) high].

[2]   Tower platform is 2'10" (0.88m) above centre of van roof and the top rail is 4'6" (1.37m) above giving an overall height of 14'4" (4.37m).

[3]   The seaward side doors have been removed and replaced by a flat panel.

[4]   Either 17 or 18 survived at Dhoon until the late 1950's

[5]   Converted from passenger car 11 in 1904.

[6]   Bodywork removed and now runs on Brush bogies.

[7]   Converted from passenger car 12 in 1903.

[8]   Converted from passenger car 13 in 1926.

[9]   These vehicles were numbered by 1918

[10]   Converted from passenger car 10 in 1926. Currently stored at Ramsey pending full restoration to 1895 condition.

[11]   Building dates are unknown but a photograph dated 1894 shows two tower wagons. The laxey fire report lists three as destroyed.

[12]   These 14 wagons are listed as in existence by 1900. Their fate is unknown although two may have been sold to Douglas Corporation for use at West Baldwin reservoir.

[13]   It is not clear whether these wagons ever entered revenue service. The rail chassis used on the MER are thought to have previously worked in Ohio. Their fate is equally unclear, the last known reference to them being in the 1901 liquidators list.

[14]   Building dates unknown but one at least must date from 1930. There may only have been two replacements for the Laxey fire victims but photographs exist of what is apparently No.3. Only one of these survives, un-numbered.

Look no wires! Isle of Man Railways' Cathie Antrobus is in charge of car 33 as it hauls its generator wagon past Keristal, on the steam railway, during a press demonstration run from Douglas to Port Erin on 3 December 1997. The overhead trolley pole has been removed from the car. This is thought to be the first time an MER car has worked away from its own line.

David Lloyd-Jones